LIBERATION AND THE COSMOS

LIBERATION AND THE COSMOS

Conversations with the Elders

Barbara A. Holmes

FORTRESS PRESS
MINNEAPOLIS

LIBERATION AND THE COSMOS
Conversations with the Elders

Copyright © 2008 Fortress Press, an imprint of Augsburg Fortress. All rights reserved. Except for brief quotations in critical articles or reviews, no part of this book may be reproduced in any manner without prior written permission from the publisher. Visit http://www.augsburgfortress.org/copyrights/contact.asp or write to Permissions, Augsburg Fortress, Box 1209, Minneapolis, MN 55440.

Cover images: John Goodman; background photo © Images.com/Corbis
Cover design: John Goodman
Book design: Michelle L. N. Cook
"Sent to Do," © 2008 Benjamin Thelonius Sanders. Used by permission.

Library of Congress Cataloging-in-Publication Data
Holmes, Barbara Ann
  Liberation and the cosmos: conversations with the elders / by Barbara A. Holmes.
    p. cm.
  Includes bibliographical references and index.
  ISBN 978-0-8006-6319-3 (alk. paper)
  1. African Americans—Religion. 2. Black theology. 3. Liberty—Religious aspects—Christianity. I. Title.
  BR563.N4H6538 2008
  305.896'073—dc22   2008029739

The paper used in this publication meets the minimum requirements of American National Standard for Information Sciences—Permanence of Paper for Printed Library Materials, ANSIZ329.48-1984.

Manufactured in the U.S.A.

12   11   10   09   08   1   2   3   4   5   6   7   8   9   10

Contents

Preface: In the Beginning ... ix
1. Introduction: The Elders Gather ... 1
2. Liberating the Idea of Freedom: ... 17
   Ain't Ya Free Yet, Why Ain't Ya Gone Free?
3. Law and Liberation: A Love Story? ... 35
   Barbara Jordan and Thurgood Marshall in Conversation
4. A Liberated and Luminous Darkness: Spirituality of Struggle ... 59
   Rosa Parks and Howard Thurman in Conversation
5. A Revolutionary Liberation: Freedom and Wholeness ... 81
   Malcolm X and Harriet Tubman in Conversation
6. Liberated Bodies, Liberated Lives: Embodying Freedom ... 99
   Audre Lorde, Fannie Lou Hamer, and George Washington Carver in Conversation
7. Killing Me, Killing You, Killing Us! Violence and Liberation ... 125
   Ida B. Wells-Barnett, Stanley Tookie Williams, and Huey P. Newton in Conversation
8. Liberation and the Art of Creative Imagination: Prophetic Expression ... 145
   Tupac Shakur, Phillis Wheatley, Langston Hughes, and Gwendolyn Brooks in Conversation
9. Beyond Mountaintops: Toward a Long-Awaited Future ... 167
   Martin Luther King Jr., Mary McLeod Bethune, W. E. B. DuBois, and Shirley Chisholm in Conversation
10. A Summing Up: Entanglements and Adjournments ... 187
    Notes ... 193
    Index ... 203

*To the storytellers, dreamkeepers, unlikely liberators,
and courageous elders and ancestors
in my personal and cosmologically extended family:*

The future is ours

Sent to Do

the deep mirrored our faces
nightingale flew on our breath
water foretold our steps
stars burned our thoughts into the sky

messages forgotten by those
shackled into staring at the groundsnow of weaving

we were the packed powder in Tubman's rounds
mamboed enchantment on stargazer's pupils
rattle shake and mama's lullaby
four-horsed father's defiant shout
freedom chanted in the clearing.

for this we come
to breathe life back into generations of slumbering limbs

the path lay before them
unfinished steps traced by our tears
sweat blood and fight song moan

we are singing kora
we are crackling revolutionary rusted string diddly bow

we are come to discuss in flesh
the coming of future things in present clearings
where the corn speaks to our seeds
where flesh fails to cloud the senses

our footsteps in the sassafras root
us who have known flesh and toil
us who turned leaning buildings into freedom temples
fueled by freedom song
writers of liberation's song

we have come
back to new beginnings
to cue freedom off pause

we have come
to make majesty of night and dream of day one
as flesh and spirit breathe in harmony
diligent toil and struggle exhale peace and justice

so we have returned
called forth by purpose
called to sound hope from pounding skin wrapped heart drums

for none shall drag our spirits away from universal will
we are come
the hurricane's first whisper
falling shoe clap of the overboard thrown

meet us in the clearing
receive the now wisdom of our words
that unfinished steps may be danced in flesh
that spirit may eternally embolden
that ancestor accompany each step.

there is much left to you to do.
hear us
see us
pray with us
believe beyond this night

we are gathered as one spirit.
meet us at the water's edge.
hear us in the clearing
rise from this riverside
go 'n shout mountain top and valley low

our spirits have come
we send you to do
we send you to do.

Benjamin Theolonius Sanders (IQ)

The movement [civil rights] is a spiritual manifestation of the continuing faith of a people who have never truly gained their rights in the nation committed by its basic law to the freedom for all.
—Derrick Bell

Preface:
In the Beginning

I once heard a story from the Native American tradition of the belief that a human being could not reach maturity without making room within herself for the immensities of the universe. That is what watching the stars helps us to do. The universe reveals itself to us, and in the hearing we are transformed.
—Judy Cannato

This is no ordinary preface. It is an invitation to engage in momentous conversations that take place in the constructive imagination of African Diasporan people. Each chapter introduces conversation partners (important figures in the African American tradition) who cross presumed boundaries between the life world as we know it and the realm of spirit and ancestry. The elders come to raise persistent and unresolved questions about the quest for freedom and its fulfillment.

We have just emerged from a century of struggle and shifting priorities, mind-boggling technological advances and cosmological discoveries, significant sociopolitical reorientations around the globe, and some confusion as to the long-term legacy of the Civil Rights movement. Like others who came of age during this era, I am intrigued by the liberation initiatives of the twentieth century, yet I am fully aware that it is too soon to objectively assess outcomes and impact. That task is for generations yet to come, but it is not too soon to frame the questions and dialogical issues that persist even as the movement fades into history.

The conversations in this book take seriously the unfinished state of the liberation project and the need for imaginative solutions that will provide the fertile soil for new initiatives. My grandmother used to say, "If you don't see a way out, stand still." After a century of activism, the desire to do something, anything, to advance the cause of freedom and to liberate those still in bondage is almost overwhelming. And yet the path forward is unclear. It seems a perfect time to regroup, to keep an eye on world events, to help a neighbor close by and imagine the future. And so this book revisits the quest for freedom using unique rhetorical tools that include the languages of cosmology, Africana thought, philosophies of liberation, and the impetus toward moral fulfillment.

I want to say a word about the discourses of science that are integral to this imaginative project. In this book, the language of cosmology serves as a friendly reference point for the exploration of a wider cultural canvas. I realize that those who study the galaxies for a living get very annoyed with theologians who dabble with their ideas without the ability to calculate the distance between one star and another. But I have no choice. I did not seek the cosmos; it found me. Because we inhabit a universe that permeates every aspect of our being, it is within our reach. In this work, cosmology is cultural and embodied, and the universe is our neighborhood, open to imaginative explorations.

In terms of methodology, this book uses the same radical creativity that inspired oppressed people throughout the ages. The only resistance to the lash of a whip or the rule of law that sustains subjugation is imagination. People move toward freedom when they imagine themselves "free," even while they are experiencing unrelenting oppression. Under such circumstances, imagination loses its "Disneyesque" characteristics and becomes a vehicle for vision and empowerment.

As a postmodern people, we no longer trust our imagination, but that's really all that we have when it comes to the configuration of freedom, the processes of liberation, or the expanse of the universe. Without imagination, our overworked minds just can't fathom the depths and illusive configurations of these ideas. And so, we begin with the rhetoric of imaginative storytelling as a conduit to realized freedoms and a liberated and beloved community.

Freedom by Any Other Name: Defining the Ethereal

Once upon a time not so long ago, a people yearned to be free. But like all things desperately desired, the essence of liberation kept escaping their grasp. I'm only the griot, the storyteller, but if you ask me, they couldn't catch the essence of it because they did not understand the differences among liberation talk, liberation laws, and liberated lives. In this book, I use the word *liberation* as a rhetorical signifier of the communal yearning of oppressed and marginalized communities to live fully emancipated lives.

I am talking about freedom, but by using the language of the liberation theology movement, I want to create a connection to the rich history and analysis that have informed the field over the last forty years. I also want to rescue the idea of "freedom" from recent self-serving, Western descriptions. Being free is not a matter of shattered chains or legislated options. There is no litmus test to sort out friends from enemies. In essence, the desire for freedom is intrinsic to the human spirit, and liberation is far too big an idea to be grounded in our personal desires and immediate social realities. Instead, freedom and liberation require the release of limitations real and imagined, internal and external, and sometimes forgiveness of the unforgivable.

Freedom is a constantly emerging state of being, reflected in the stories and questions of its seekers, in the liberation talk of its leaders, in the creative interplay of the already/not yet, and in the particularity of personal and communal experiences writ large and told often. *Liberation* and *freedom* are words that point toward the same goals but may have diverging meanings and cultural history. A full discussion of definitional tensions and interfaces can be found in chapter 2.

But I digress. According to the stories told, the imaginative rhetoric of iconic liberation leaders and freedom fighters inspired an entire generation during their lives. However, after their deaths, the slogans and memories of their activism seemed too flimsy to serve as a foundation for the survival of the African Diasporan community. Is it any wonder that in the silence that followed the glory days of the Civil Rights movement, people became impatient? They wanted more than stories and memories.

They wanted answers and tangible proof of their freedom. And so it was that some who sought liberation chose earthbound ideas instead of living stories. Although they knew better, they ignored the gatherings in the meeting places where living knowledge was passed around kitchen tables and in barbershops, beauty parlors, and choir meetings. Instead, they entrusted established civil and religious systems with the task of enforcing and protecting the fragile spark of liberation.

Not a good idea. The systems chosen to guard the flame were in transition. They presumed that the spark would continue to flicker while they attended to their own urgent business. Meanwhile, translators of the meaning of this spark of freedom were few and far between. Because liberation was still only a glint, a hope, a spark, when the elders transitioned to the life beyond life, and because attainment took so much energy and time, the next generation could not sustain a clear vision of what it means to be free. So they said to one another, sometimes in whispers and other times in shouts, "If we can just figure out what liberation looks like, how it smells, what it feels like, we'll be able to cage or clutch it so that we can pass it on to our children." It took awhile to realize that this would never work, because liberation comes with wings and knows how to fly. Liberation fades under intense scrutiny but flourishes and expands in the stories told by its seekers.

And so things went awry. Now I'm not saying that the liberation project of the twentieth century was a failure. On more than one occasion, those who sought freedom and liberation from bondage came very close to realizing their dream. On several occasions, they caught a glimpse of its shadow, but a shadow is only an impression and they needed more. Liberation came and went in spurts during the twentieth century. It might have been easier to catch hold of a greased pig. Eventually, seekers realized that the permanence of liberation gained, domesticated, and final would continue to elude them. And yet, they truly believed that eventually things would work out. This belief was seeded in a resolute refusal to give up.

But here is the thing: giving up is not an event that occurs with trumpets blaring and announcers pronouncing. Giving up is gradual. Little by little, even as they declared their steadfast commitment to liberation they redefined the dream. Soon seekers were chasing after phantoms and imitating others who seemed to be free. When their patience wore thin, they tried to force freedom to show itself by declaring its appearance where it was not and by describing its attributes in the most incongruous of ways. And so, liberation became synonymous with education, purchasing power, and consumer lust.

Ultimately, the definition of liberation in Holy Scripture and the definition that served as the core principle of civil religion got entangled to the point that liberation became synonymous with the pursuit of personal happiness and wealth by any means necessary, with the satisfaction of individual needs, private entitlements, and corporate and political control of the public sphere. Accordingly, these pursuits encouraged the neglect of communal responsibility.

It was a last straw for the God of liberation, whose revelation occurs within the context of a holy community (God, Son, Holy Spirit), within the visible and invisible elements of a creative and self-reflective universe, and in the mimetic rhythm of life captured in the resonance of every heartbeat. When ancient people gazed at ancient stars and shared stories around glowing fires, they imagined themselves free. But no one looked up anymore. Instead of finding clues to their ultimate liberation in the cosmos, the people who had not been a people were lost in a morass of unenlightened self-interest.

It was time for the fog to lift. It was time for the idea of freedom to rise from the mire of political and legal entanglements to expose its embeddedness in the radical creativity and regenerative steadfastness of the cosmos. A meeting would be called of ancestors and elders, of predecessors and long-gone Aunties, to remind generations still longing to be free, of the liberation seeded from the beginning of time in the cosmos.

For me, writing is both a privilege and a necessity. Through my books I amplify the wisdom of generations long gone, and teachers, mentors, and colleagues still journeying with me. Most writers spin worlds from their laptops in relative solitude, but they do not write alone. Always, the voices of friends, family, predecessors, teachers, and colleagues provide the engagement and testing ground for the work. I am grateful for the guidance of editor-in-chief Michael West and others (Julie O'Brien and Susan Johnson). Creative ideas become good books because of their insightful suggestions. Also, I am grateful to my teachers. Howard Harrod introduced me to phenomenology and the world of predecessors in the work of Alfred Schutz, whose theories provide a foundation for this book. Schutz argued that, "The line separating present social reality from the world of predecessors is fluid. Simply by looking at them in a different light, I can interpret my memories of people I have known directly or indirectly as if these memories belonged to the world of my predecessors."

Victor Anderson introduced me to the philosophy of religion, the relevance of cultural studies to ethics, and Hans Georg Gadamer's *Truth and Method*. The influences of those reflective and reflexive dialogical methodologies are also evident in this book. Lewis V. Baldwin taught me to explore the contributions of Africans in the Diaspora, thereby providing a historical grounding for my metaphysical and imaginative theories. Marcia Y. Riggs reminded me of the priority of justice and taught a systems approach to dismantling the "isms."

Finally, I am grateful to my mother for sharing her life with me during her ninth decade. She is a living ancestor reminding me of where I've been and how joyful the future can be. For this and much more, I am grateful.

Storytelling promises to make meaning out of raw experiences; to transcend suffering; to offer warnings, advice and other guidance; to provide a means of traveling beyond the personal; and to provide inspiration, entertainment, and new frames of reference to both tellers and listeners.
—Amy Shuman

1. Introduction: The Elders Gather

These are the storylines of our existence. They are not mere wisps of non-inert energy. They are weighty. We embody them. . . . We carry them with us as if we are walking around with a mass on our shoulders. We have each bought and sold ourselves on a certain story. . . . Any idea that runs counter to our story even if it could improve our situation, we quickly push down into a safe comfortable zone of unthinkability.

In order to break through, in order to have a new experience, a shamanic awakening, we have to break free of the illusion that we are separate from anything else, and in particular, free of what we desire. . . . We must tap into the imagined realm to create a new storyline.
—Fred Alan Wolf

The faces gathered around the meeting table were familiar. There was Martin, of course, Rosa, Ida B., and Marcus Garvey. El-Hajj Malik El-Shabazz, or Malcolm X, as he was known for one period of his life on earth, was pacing, waiting for W. E. B. and Fannie Lou to arrive.[1] On the other side, they saw one another more often than they had during their lifetimes, but meetings were rare, called only during periods of crisis.

The elders considered the status of the liberation projects to be at a critical juncture, and so they gathered. When Fannie Lou arrived, she said to no one in particular, "Aren't they sick and tired of being sick and tired yet?" "That really isn't the point at all," W. E. B. interjected. "You can get so used to being sick that sick seems normal. That's why I had to leave the West."

"And that's why I advocated that we all leave," said Malcolm. "You weren't the first," Garvey said, smiling. "I even had boats ready."

George Washington Carver said, "Look, I told you folks that the talented tenth needed to know how to grow a few collard greens. Now you've got a bunch of educated zombies, who don't even realize that they haven't left Egypt."[2] At this point, Barbara Jordan chimed in, "Would someone please tell me what collards have to do with liberation?"

When Martin spoke, everyone hushed the chatter and cross talk to hear what he had to say. "God's timing is crucial in all things," he said. Martin still spoke with the rhythmic power of a Baptist preacher, but there was a certain ease in his voice, presumably because he no longer had the responsibility of leading a reluctant nation toward its higher good. Coretta sat on his right side, one hand lightly touching his arm. Although marriage did not exist in this place, they chose to be together most of the time. "Look," he said, "we need to tell our stories to another generation on the other side of the veil. So, let's hold off on any further conversation until the Writer gets here."

"Who is the Writer?" Langston Hughes sniffed. "I'm here, aren't I?" Martin laughed with the joy of one who is truly free, "Of course you are *The* Writer, but we need someone who can publish our thoughts on earth. She or he need not be as famous or as talented as you are, just willing, and I think that I know who is available. The woman that I have in mind isn't worried about crossing disciplinary lines and is not constrained by academic rules. The fact that she already writes weird stuff makes her perfect for this task."

They were talking about me. How I happened to get involved with this project is a story that also needs to be told. But I will say more about the assignment later. What I have described to you is only the preliminary conversation of those at the meeting. By now, you are probably curious about the location of this event. For ease of reference, I named this betwixt and between geo-spiritual space "the other side." I realize that in a world where there are really no objective ups and downs, there are also no sides. But because the leaders who are the focus of this narrative are on the other side of the life-death continuum, it helps conceptually to designate the realm of the departed with terms that have some familiarity.

I realize that this is an inadequate explanation, but I can only tell you what I know. I know that African foremothers and forefathers would have referred to the assembled leaders as ancestors and that the place would be understood to be "beyond the veil." Although some folks use a very narrow definition of the word *ancestor*, I use the word as an indicator of legacy and interconnections. The ancestors are elders who pour their lives into the community as a libation of love and commitment. They live and die well, and when they transition, they do so in full connection with an engaged community.

Thereafter, they dwell in the spaces carved out by our spiritual and cultural expectations. They may be in another life dimension, but they connect with us in dreams, in memories, and in stories. In our family on the Gullah side, we expect ancestors to come and help us cross over to the other side when it is time for us to leave this earth behind. And, perhaps because we expect it, they do. A few of my North American Indian, African, and Japanese friends tell me that the same thing happens in their families.[3]

The stories reveal a promise that the community will continue beyond the breath of one individual and that all transitions will be well attended by relatives from the other side. This is a cosmology of connection that values but also transcends cultural contexts; life is considered to be a continuum of transitions, ruptures, and returns. Those who admit that the "ordinary" is punctuated by the ineffable cherish those indescribable and nonrational events as an enigmatic but welcome gift. The fact that I grew up in a family that included the presumptions of transcendence and the unseen in our everyday lives has affected my journey in powerful ways. Even before I was assigned this unusual writing task, my life was punctuated with stories and connections bequeathed by my elders.

The end result is that I know that I am not alone. I am connected to the past and the future by the ligatures of well-lived lives, the mysteries of "beyondness," and the memories and narratives that lovingly bind and support me. While I hope that when I die, one of the elders in my family who have crossed over to the realm of the ancestors will be at my bedside, I certainly did not expect contact prior to that time. And yet here I am, hearing from liberation leaders I have never personally met. They are also my elders as certainly as if they occupied a branch of my family tree. They have bequeathed to all of us a legacy of resolve, resistance, and spiritual expansiveness.

As I ponder these things, I realize that the space around me is getting very quiet. People continue to assemble, but their movements seem meditative and slow. It is difficult to determine whether this is part of my own dream rhythm or of theirs. But the slowing and silence give me a chance to "see" the configuration of the gathering space. It seemed to be circular, completely open but intimate. But this may be a space of my own making, one I have imagined as a trustworthy indicator that there will be no hierarchical power struggles or self-importance. Because we are on "the other side," the space does not get overcrowded, no matter how many luminaries appear. I can only guess that this is a space-time dimensional phenomenon that I don't need to understand. It occurs to me that I am still struggling to make this experience conform to my rationalist expectations, and of course it won't oblige.

One of my persistent expectations is that justice seeking and freedom fighting are serious business and that liberation is jaw-clenching, teeth-gnashing work. I find it odd that this cloud of witnesses seems particularly lighthearted, and I marvel at their congeniality. I can't help but wonder about what happens to all of the ego-sustaining practices that we nurture during the life journey? Apparently, during the transition from one realm to another, wisdom, memory, and the recognition of loved ones remain but prestige and power are left behind. Those who have come to share their stories have joy that manifests as insight that is not strident and prophetic wisdom that is not arrogant. Now, these liberation leaders want to share their wisdom in an effort to awaken and inspire another generation.

It so happens that this meeting is being called by liberation leaders of African Diasporan descent. Although they have a particular love for the African Diasporan people who they served on earth, they consider their progeny to be all who struggle for justice and liberation, without regard to race, class, or gender/sexuality. The meeting is for the benefit of all who love justice, but the intent is to address the particularity of the liberation project in an African American context.

Liberation is an egalitarian event that can't be confined to one community or another. The quest for liberation transcends the desires, failures, and intentions of its seekers even as it remains integral to their stories of captivity and deliverance. At this meeting, many will gather, some famous in life, others unknown local leaders, some from other cultural contexts, some martyrs whose names have long been forgotten. There will be abolitionists and freedom fighters of every ilk. It is the first of many wisdom circles. What distinguishes the gathering is that the participants have all made their transition to the spirit realm.

From what I can understand at this point, they are gathering because the justice movements have been derailed and new generations are looking in the wrong places for freedom. It wasn't apparent at first, but after the liberation movements faded, after many of the elders of the Civil Rights movement crossed over, after the enticements of the marketplace that were electronically dangled before the bleary eyes of the oppressed "changed the object of their desires," everyone went to the mall.[4] Those who had struggled and protested to achieve liberation re-enslaved themselves to their credit cards and became debtors making bricks without straw or health care. Now there was only the occasional reality show to distract them from their plight.

It now seems apparent that "the powers that be" found slavery to be such a lucrative enterprise and such a psychologically beneficial tool in the control of human populations that they now offer it (irrespective of race) as an equal opportunity assignment around the globe. Although the oppressed did seem to have gained some access to wealth and power in the twenty-first century, this inclusion in the market economy was no indicator of wellness. The trickle-down theory did not work any better with rich rappers than it had with wealthy plantation owners. While the market economy allowed a select few to reap benefits, that same system exploited the remainder of the population and the labor of people in emerging nations around the globe.

The elders also heard the cries of the poor, whose plight seemed even worse than before the justice movements, and there were new scapegoats in the twenty-first century. The word *terror* kept coming up. The word seemed to have been ripped from its ordinary context to serve as a rhetorical marker for fear itself. As bizarre as it may seem, "fear not" Christians aligned themselves with the powers wreaking preemptive havoc throughout the world. Religion in the West at the end of the twentieth century became a stalwart ally of the domination systems, concerning themselves with pleasing their congregational audiences and ensuring a steady flow of income.

Slowly but surely, the language of prophecy and the scandal of the Gospel was displaced by feel-good rhetoric. But even more disturbing was the use of religious discourse as a cultural club and a weapon of choice all around the globe. No one knew how to revive the communal efforts toward liberation, former alliances fractured on the anvil of sexuality, gender, and ethnicity. Soon, those who wanted to leave Egypt were confused and speechless. They no longer picked cotton on plantations; instead, they flipped burgers in fast-food joints to pay for cell phones and cable television. Eventually, the groans of another generation toiling with their eyes cast downward caught the attention of the ancestors.

With regard to race relations, at the dawn of the twenty-first century things looked great on the surface. An African American man was a presidential candidate, and several decades of integration seemed to have changed the rigid lines of social demarcation. There were no more sit-ins, no more protests, but now violence seemed as internal as it was external. When it manifested, it was not the noose of the lynch mob but the callous disregard of neighbors, friends, and family members that threatened the safety and well-being of the community. Killing was now integral to life in community—an act that was intimately configured and confined within the boundaries of the neighborhood or the home.

The end result seemed obvious to the ancestors: the survival of a diasporan people who had overcome slavery, Jim Crow, economic impoverishment, and violence was now endangered. Clearly, survival is not a solid foundation for the advancement of liberation initiatives because survival often masks unseen stress fractures that threaten both individuals and communities. Moreover, oppression breaks the spirits and bodies of its victims. It inscribes on the souls of "survivors" a story of suffering that is difficult to escape even when outward circumstances change for the better. The fact that the pressure lessens because of societal shifts or legal mandates does not address the damage done.

Ostensible inclusion in political or social systems is not a reliable indicator of a successful ascent from oppression, since the sickness of oppression is contagious and mimetic. Its victims become interns. During the process of their own degradation, they learn how to be oppressors in new and innovative ways. We see it continually but refuse to engage this socio-spiritual aspect of what seems to be very mundane. The worker in the plant who is abused by the corporate system takes it out on his family, the parent turns on the child, and the economic systems that relentlessly abuse labor in emerging countries eventually lose their moorings to all that is just.

Without healing, endangered people contribute to their own spiritual and cultural extinction. Self-destruction takes many forms. It can come in the refusal to care for the body, submitting it to casual "intimacies," starving it with fast food. It can come in the enslavement of the mind and devotion to the religion of consumerism. It can convince nearly free people to kill one another over turf and ego. The stories of destruction abound, and although they must be remembered to avoid repetition, the future does not lie in the retelling of such tales. Much more is needed.

Stories of connection and cosmological awareness are vital. Future generations will rely on the story lines that are also lifelines of perseverance, grace, and mercy. These stories remind us that no matter what circumstances we endure, there are always opportunities for unexpected deliverance and encounters with a narrative God. The cosmos offers imprints of creativity and hope based on the potential of the future and the generative wisdom of the past.

I was beginning to understand the importance of this meeting and the conversations that would follow. Liberation takes on an entirely different meaning when the psyche, the body, and the connections to something greater than the self are factored into the equation. Liberation can't be stamped on the foreheads of its seekers; rather, its contours emerge as the stories of failure and success are told and retold. The ancestors were prepared to whisper the wisdom of the ages into the spirits of another generation. And so, it was determined that those who had sacrificed so much in the struggle for justice would gather.

When I was first contacted to sit in on this meeting and to write what I heard, I wasn't certain that it was anything but a strange dream. But a series of these reveries that followed in logical and sequential order convinced me that this was a bit more than the usual "too much pizza" sleep disturbance. Especially disconcerting was the fact that some contact occurred when I was awake and in the middle of a business conversation. I would find myself planning a fall schedule of classes with the academic team at my seminary when a conversation in midsentence would begin with Rosa or Malcolm. It was disconcerting to say the least. I am grateful that my colleagues would simply look at one another and occasionally murmur, "She's writing books in her head again," whenever I was present in the body but absent in every other way.

But my discomfort and lurching acceptance of these odd encounters convinced the elders that I indeed needed a designated contact person to answer my questions and to prepare me for what would occur. They chose a martyr named Sarah, who had been one of a number of unknown murdered resisters in Mississippi during the years that led up to the Civil Rights movement. She was a composite of all the unknown women who gave their lives in the struggle for liberation, but she also had been a flesh and blood woman who was one of Fannie Lou's forerunners. Now her task was to help me understand the ethereal procedures that would lead up to the meeting.

As I understood it, Sarah would try to answer my questions in ways that I could understand, given my embodied and thereby limited access to realities beyond the veil. And I did have questions. During one particularly vivid encounter, I asked Sarah whether I should be interfacing with persons on the other side. It was one of those "ought" questions that is always on the tip of an ethicist's tongue; it was also a question that arose from my current social location in the Bible Belt, where some prevalent Christian interpretations of God's prohibitions prevailed.

Although I approach Holy Scripture as a living repository of wisdom that invites dialogue and critique, I know enough Bible to know that God was pretty annoyed with King Saul for conjuring Samuel. When Sarah heard my reasoning, she laughed raucously. Then she assured me that the linkages of love and commitment survive the transitions of life and death and that conjuring would not be necessary. When I balked, she just winked.

Sarah You're worried about conjuring? Aren't your people from the Gullah islands?

Just what you need, I thought to myself, a smart-aleck ancestor. Then Sarah offered an example that helped me to understand the rules of engagement. She reminded me of the vibrancy of the stars in an Arizona night sky. I have seen them and know that they are like precious jewels strewn across a multidimensional darkness. The stars are always there, but I cannot see them as well in the cities where I live because artificial lights obscure their brilliance.

Sarah It's the same thing. Conjuring isn't necessary. The spirit realm is always within reach whether you can see it or not.

It took a minute for me to realize just how arrogant it was to assume that earthbound souls could drag elders anywhere.

Sarah The ancestors are always moving from realm to realm, escorting newborns and helping people to cross over to the life after life, but you can't see beyond the limits of life with earthbound eyes. On earth, when you walk by the light of your own ego--stoked fires, your lights—though dim and unreliable—obscure the light and wisdom from the stars. We are the stars in your midst.

When I continued to look troubled, Sarah reminded me that many ancient religions openly acknowledge predecessors.

Sarah Christianity hides its ancestors in plain view. Those familiar with the Bible know that Jesus had a very public conversation with ancestors in full view of chosen disciples.

Now it was my turn to laugh when she added "and he didn't get into trouble, so you're probably all right." She's right. We choose safe words and images like *prayer* and *transfiguration* to soothe our discomfort with ancestor contacts that require the crossing of dimensions.

Writer Sarah, you keep using the word *ancestor*. I thought that this term was reserved for special people with special powers to aid the living.

Sarah Well, it's a bit more than that. Ancestors within the context of African Diasporan legacies are those family members who have poured out their lives for the good of the family. In life, they live well and for others. Although they are human and fail often, they are seldom deluded by the distractions of ego or the desire for earthly acquisitions. They also transition well to the other side and continue their intercession and prayers for the living. Most of the African Diaspora originated in West Africa, and so a view of their cosmology is relevant.

> In addition to the creator and the orishas, the West African sacred cosmos includes both familial and tribal spirits of those who no longer live within the physical dimension. These are the ancestors.[5]

Writer But none of these people were in my family.

Sarah Daughter, your family is bigger than you think. The linkages of care and connection far exceed genetic progeny. There are spiritual mentors, visionaries, and prophets who impact your life throughout the generations.

> The ancestors are familial in the sense that they may be the "elevated" spirits of a particular family. They also may become tribal when an entire clan, tribe or nation venerates an extraordinary human being (as for example, the cases of Martin Luther King, Jr., or Malcolm X within the African American religio-cultural experience).[6]

Writer Will they all be African American?

Sarah Those distinctions don't exist here. The reality of the earth experience is valued and honored, but the spirits that you will encounter are spirits, no more no less. So wisdom may be offered by folks who are not directly from the African Diasporan experience. This conversation is being watched closely as a model for others to speak to their own cultural groups.

Writer I think that I understand, but what shall I call them?

Sarah They don't care whether you call them ancestors or elders or by their earth-side names. Now that they exist in another dimension, they have a different view of identity. But they'll tell you about that themselves.

Speaking of dimensions, I wondered about how the dimension of time would work during this encounter.

Writer Will I be 110 years old when I wake from my reverie, because if that is the case, all bets are off.

Sarah You certainly do worry a lot. I can't tell you everything, and even if I could, you wouldn't understand it. But I can say this: time does not have the same meaning in eternity as it does on earth. Certainly, it is a dimension that can conflate and expand, but it is not the whip-wielding, punctuality-enforcing dominatrix that you know in your realm. That is the way of the world. Here, a minute is like a thousand years, and vice versa. Please leave the details to us, and don't forget that the questions will drive the conversation.

It was settled. I would be an observer and an interviewer of clusters of designated spokespersons. My task was to record the observations and suggestions of the elders and to help to frame the interlocutory questions that would tease out their wisdom. Each conversation would be focused around a key question and would address a specific issue in the community or speak to the needs of a particular group.

Although there were other ways to get this message out—such as skywriting and booming voices out of nowhere or collective dream messages to large groups of people (which I much preferred)—Sarah patiently explained that the most effective spiritual messages were those that were whispered rather than roared. So it was decided that the elders' commentary to this and future generations would come in the form of a small book that I would write. Some would read it; some would not. Some would consider it a silly fictional piece, without academic weight. But for those who had ears to hear, the word would be received and the progeny of these great leaders would learn about the legacy of liberation and its future.

Sarah Are you ready? Go to bed early tonight, and when you wake up, the stories will remain with you.

Since readiness has never been a prerequisite for my willingness to venture down dimly lit spiritual paths or to make risky choices, I decided that, ready or not, it was time for the meeting to begin.

What you will read in the next several chapters are the hastily jotted notes that I took and assembled after one of the most restful nights I've ever had. I awoke refreshed and with a sense of purpose and commitment to the project. The outline and order of the conversations are mine. I really can't blame the elders for any errors that the reader may find. Because this information has been filtered through the sensibilities of a willing but human writer, flaws are inevitable. In this regard, I've been assured by the elders that "order is as order does." This is a slightly reworded version of an old African Diasporan proverb that simply reminds me to listen, to be still and accept the unseen order that often emerges out of chaos. Always, the discussions express the rhetorical and descriptive context of liberation and its meaning using key words that connect human and cosmological narratives.

The Questions: Talking about the Elephants in the Room

I will describe the chapters to the best of my ability but presume that the conversations can't be predicted. Here is the plan, whether we stick to it or not. Each chapter situates the elders and their individual life journey and spiritual intent within the context of a themed conversation. An overarching question informs the discussion. The questions raised are not definitive; they are provisional, processive, evocative, and interlocutory.[7] They focus on unaddressed issues that tend to arise whenever freedom and liberation are discussed.

Each chapter begins with brief biographies of the elders and ends with a dialogical analysis of the conversation and suggestions for further reader reflection. The chapters unfold as follows: This and the following introductory chapter offer a framework for the ensuing conversations as well as a premeeting discussion of the meaning of freedom and liberation within the context of the story of the universe. Chapter 3, "Law and Liberation: A Love Story?" records the conversation of Barbara Jordan and Thurgood Marshall regarding the relationship among law, politics, and liberation as they relate to the science of indeterminism. The question under consideration is how the laws of society and the cosmos impact the quest for liberation.

Chapter 4, "A Liberated and Luminous Darkness," invites a contemplative view of liberation. Rosa Parks and Howard Thurman are the conversation partners who revisit Thurman's idea of "luminous darkness" in conjunction with a liberationist view of dark matter and dark energy. Question: Does liberation have a discernible spiritual interiority? Chapter 5, "A Revolutionary Liberation," features a conversation with Malcolm X and Harriet Tubman about resistance, community, and a cultural interpretation of holographic effects. Question: Is there a relationship among revolution, cosmology, and wholeness? In chapter 6, "Liberated Bodies, Liberated Lives," Audre Lorde, Fannie Lou Hamer, and George Washington Carver consider the relationship among faith, health, and the well-being of besieged communities. Question: Is there a connection among body wisdom, generational restoration, and the science of posthumanism?

Chapter 7, "Killing Me, Killing You, Killing Us!" speaks—through the voices of Ida B. Wells, Huey P. Newton, and Stanley Tookie Williams—to the issue of state-sanctioned and random violence. Topics include violence and communal healing, while the cosmological orientation includes chaos theory and nonlocality. Question: How can we consider the quest for liberation within the context of a social and personal "cosmology of violence"? Chapter 8, "Liberation and the Art of Creative Imagination," includes a conversation with Tupac Shakur, Phillis Wheatley, Langston Hughes, and Gwendolyn Brooks. Question: What is the role of the prophetic artist in the struggle for liberation?

Finally, in chapter 9, "Beyond Mountaintops: Toward a Long-Awaited Future," three liberationist leaders—Martin Luther King Jr., Mary McLeod Bethune, and W. E. B. DuBois—talk about multiple dimensions and the future. Question: What is the future shape and impetus of liberation? In each of these conversations, the odd thing for me is that the science seems so integral to the discussions. Underlying all of the ideas and questions that emerge in each chapter is the interconnectedness of the universe and the human enterprise.

I summarize what I can in chapter 10 and hope for a postscript from the other side. I have much to learn, but I do know that conversations about freedom in the twenty-first century are difficult to say the least. Those who feel that not much has changed with regard to freedom and liberation don't want to hear that Oprah Winfrey, Condoleezza Rice, and Barack Obama prove that African Diasporan people have arrived and that they should stop complaining. People of color in non-Western global communities wonder why the conversation seldom includes them and why justice "talk" never confronts the economic inequities inherent in international debt calculations.

I compare this equal opportunity discomfort around the issue of liberation to the attempt to have a meaningful conversation in a small room with a herd of unhappy elephants. The elephants are the subjects that no one wants to address directly. Liberation theologians have been edging around their swinging trunks and huge feet, trying to cobble together an approach to liberation philosophy that is attentive to history and efficacious for the future. We want to offer an honest appraisal of the state of this life-giving quest, but there are so many things that we don't want to talk about. One elephant in the room would be difficult enough, but I am identifying ten, hereafter referred to as E-1 through E-10. There may be even more, but I will leave further identifications to the reader.

E-1. *What does it mean to be free?* How do we evaluate the freedom of individuals and groups? If and when freedom is obtained, how will we know it and who is entitled to the benefits? Are power, choice, wealth, and autonomy reliable indicators of individual or collective freedom? What are the tensions between the freedom of global capitalists and the liberation of indigenous and base communities?

E-2. *What does freedom mean in a cosmological context?* I am raising again the question posed in *Race and the Cosmos:* Can we align social constructs of liberation and justice with emerging cosmic realities?[8] The presumption in this book is that liberation is more than an outcome; it is, indeed, a cosmology of interrelated constellations of hope and desire, of seen and unseen linkages. "Emerging views of a cosmos that is characterized by holism, indeterminacy, and interconnections have the potential to empower people from the two-thirds world in ways that liberation theology has not."[9] This view of the goals and processes of liberation expand the options for fulfillment of generational dreams.

According to the known laws of cosmology, the universe is delicately balanced to support life. How does this incredible scientific fact affect prevailing concepts of free will? What did Martin Luther King Jr. mean when he said "the moral arc of the universe bends at the elbow of justice?" How do we describe a cosmology of liberation, and how does a cosmic view of global freedom struggles move us toward solidarity?

E-3. *What does law have to do with freedom?* How do we balance our ideal configurations of freedom in an ordered society with the claims of those who cross borders (legally or illegally) to participate in that society? Can we identify instances when the law was or was not on the side of the poor and the oppressed? How does the definition of the word *freedom* differ in the twenty-first century from the ideal of the founders of the United States and from the objectives of the Civil Rights movement?

Since the Civil Rights movement, laws have been considered the source and sustainers of liberation, yet the laws passed during the 1960s settled only the most basic disputes about human rights and resolved none of our suspicions about one another. The law supported segregation until the sheer weight of degradation pulled legal precedents apart at the seams. Why is it difficult to sustain a conversation about unequal law enforcement that leaves marginalized communities at risk and under siege?

E-4. *Can we identify spiritual characteristics of freedom?* Are there spiritual practices that support a consciousness conducive to freedom? Can we identify specific aspects of at least three faith traditions that support the idea of liberation? What is the connection between historic struggles for liberation and the spiritual journey of embattled communities? Is the idea of liberation capable of serving the needs of civil religion and Christian values? What are the differences, if any?

E-5. *Do liberation initiatives require a revolution of values?* What can be compromised and what must be resisted with regard to liberation initiatives in a free society? Are there basic liberation values and moral imperatives that can be consistently applied on behalf of oppressed communities around the globe? When someone says, "They hate our freedom," what does that person mean?

E-6. *How is freedom embodied?* Can activists agree on what liberation means within the context of race, class, gender, and sexuality divisions? What does it mean to "be free" in the body or to be "liberated" in an embodied way? How does the current view of human liberation comport with the abuse and domination of nature? Can we map tributaries of self-care and liberation in our own bodies? How does our Christian understanding of "light" and "darkness" contribute to or hinder the liberation of people of color around the world?

E-7. *What is the relationship between violence and liberation?* Considering a history of successful liberation initiatives in Africa (South Africa), the United States (the Civil Rights movement), and Europe (Berlin), can we identify common elements in each of those successes? Is it possible to determine what came first: the desire for freedom, revolution, a liberated consciousness, or changed laws and social structures? Is torture "new wave" lynching? Can it ever be justified when used to advance the cause of a nation's interest in "freedom"? How do our prison system and the death penalty support or undermine our faith understanding of liberation and forgiveness?

E-8. *What is the role of prophetic art and artists in the quest for liberation?* Is there a connection between art and social transformation or liberation? Can art liberate oppressed communities? Does hip-hop/rap have a prophetic dimension? Is there a current equivalence to the liberation aspects of the Black Arts movement of the 1960s? Can art address questions of violence, economic empowerment, and justice?

E-9. *What will liberation look like in the future?* Can we connect familiar stories of oppression (African American, American Indian) with new "coming of age" liberation stories (Latino/a, Asian, African)? Do we still need or want the community called beloved, and who will be included or excluded? Will virtual communities supplant the vision of embodied unity? Does the age of technology change the notion of freedom, its objectives, and its leadership?

E-10. *Who will lead?* Is the era of luminaries over? Can we create a beloved community without a designated leader? What does leadership toward a liberated existence look like in the twenty-first century?[10]

The questions have been framed, the elephants in the room have been identified, and now I have the feeling that folks are waiting for me. It's time to begin the journey.

Oh Freedom, Oh Freedom,
Oh Freedom over me, and before I'd
be a slave, I'd be buried in my grave
and go home to my God and be free!
    —Negro spiritual

2. Liberating the Idea of Freedom: "Ain't Ya Free Yet, Why Ain't Ya Gone Free?"[1]

A human being is part of a whole, called by us the "Universe," a part limited in time and space. He [or she] experiences . . . thoughts and feelings, as something separated from the rest—a kind of optical delusion of . . . consciousness. This delusion is a kind of prison for us, restricting us to our personal desires and to affection for a few persons nearest us. Our task must be to free ourselves from this prison by widening our circles of compassion to embrace all living creatures and the whole of nature in its beauty. The true value of a human being is determined by the measure and the sense in which they have obtained liberation from the self. We shall require a substantially new manner of thinking if humanity is to survive.
    —Albert Einstein

I was ready to start the interviews, a little nervous but ready. But I needed to clarify the theme, and I wondered about the key questions that would frame the discussions. First, I wanted to understand clearly the meaning and usage of the words "*liberation*" and "*freedom*." For the purposes of these conversations, are the words being used interchangeably? Second, I wanted to address the issues that arise whenever the potential for sustainable liberations exists.

Finally, I knew that the conversations would address the black liberation movement from a cosmological perspective, but I didn't really know what that meant. What would it mean to have a "cosmology of liberation" linked to constellations of hope and solidarity? To the best of my knowledge, none of the leaders, artists, and activists who were assembling had any expertise in astronomy or quantum physics. So how would we connect the very real work of liberation to cosmology? If I could put these three issues into context, then the conversations would probably have more meaning for the reader.

Freedom by Any Other Name: Defining the Ethereal

How do you define an idea like freedom that is expressed in the natural but also rooted in the spirit? There is no doubt that those denied freedom have tangible proof of their bondage etched into body and spirit, but release from chains does not necessarily imply that those liberated have the inner strength required to sustain those gains. Liberation is a multidimensional process. If it is tied to immediate oppressions without reference to the opaque and unthinkable future, then one can be certain that those gains will not fit the needs of generations yet to come.

Justice gained in cotton fields may not translate into corporate or technological lifestyles, even though a corporate cubicle, an academic tenure track, and a row of cotton may have very similar productivity expectations. The struggle for the right to vote, once gained, does not address the root of the problem when those votes are allowed but are discounted by electronic voting machines or "hanging chads."

The quest for liberation/freedom in any social setting must have at its root a spark that is replenished by human, divine, and cosmological resources. Although there may be an initiating incident at one point in history—for example, Katrina or Jena 6—the underlying ethical and social impetus toward liberation must be more expansive than the incident. The courage to confront the systematic powers of oppression requires a sense of connection to that which emanates from within and beyond human limits.

I am certain that the idea of liberation was understood by all who participated in the Civil Rights movement. And yet, at the beginning of the twenty-first century, after all of those years of struggle, sit-ins, court battles, and assassinations, some in the community wonder, "Ain't we free yet, why ain't we gone free?" It turns out that integration is not the same as liberation and that striving will not ensure freedom.

Philosophies and Cosmologies of Liberation and Freedom

The philosophies of liberation that inform this work were constructed by marginalized people in opposition to presumptions of inferiority and nonbeing that pervaded some of the dominant Western philosophies. Against all odds, scholars, local activists, and religious leaders began to share ideas and write about the state of liberation that they envisioned. Enrique Dussel, Daisy Machado, James Cone, Katie Cannon, Kwok Pui-Lan, Anthony Pinn, Angela Davis, Justo Gonzales, Victor Anderson, Emilie Townes, Gustavo Guitierrez, Cornel West, Marcia Y. Riggs, William R. Jones, Charles Long, and Ada Maria Isasis Diaz, to name just a few, approached the systems of oppression with persistence and scholarly incisiveness. Although these individuals used different perspectives drawn from cultural, and socioreligious experiences, they all "spoke truth to power."

In African Diasporan contexts, ideas about freedom and liberation seemed to be drawn from a view of the world that is unbifurcated, whole, and populated by the living and the dead. Albert Raboteau says this about the West African mind-set that informed this view of liberation: "African spirituality does not dichotomize body and Spirit, but views the human being as embodied spirit and inspirited body, so that the whole person—body and spirit—is involved in the worship of God."[2] And always, worship included reinforcement of the belief that liberation could come about through the agency of freedom-oriented people and interventions from the spirit realm. African Diasporan philosophies of liberation include the natural world and do not depend on extreme categories of human good or evil. Instead, they emerge from immediate cultural realities and creative adaptations from various sources.

Contributions from the legacy of Plato and Aristotle include descriptions of freedom as "inseparable from the political arena [and] . . . contingent upon the greater public world."[3] The advent of Christianity adds to this concept an understanding of freedom as liberation from sin.[4] It is clear that ideas about freedom often require a negotiation of claims. What happens when a rancher on the border of Mexico asserts her right of freedom to protect her land from trespassers in opposition to the freedom of impoverished Mexicans who are crossing the border to work?

One thing is certain: freedom can't be described in the same way for all people or all times. What it means to be free or liberated from bondage ranges from personal choice and collective power to the orientation of consciousness. There are as many expressions of freedom as there are options for dominance.

There is no way of life that is the free way: freedom implies diversity, novelty, individual uniqueness. . . . Liberation is not won with a single act of rebellion or violent impulse, nor is it given to people by the leadership. It is built up from below through the collaborative struggle of the people. A successful armed struggle only removes an intolerable obstacle; it does not in itself usher in liberation, just as the act of leaving Egypt did not mean immediately being in the Promised Land. The long period of "wandering in the wilderness" cannot be avoided, a period in which there will be outside attack, internal rebellion, the temptation to return to the past and to shift to the worship of false gods. It is only in the course of such a journey that liberation can be constructed.[5]

Given the fluid and processive nature of liberation/freedom, it seems appropriate to consider definitions at this point. Most dictionaries define *liberation* as release from something, while *freedom* refers to the absence of coercion or restraint and to a state of being that defines the nature of human existence. For African Americans, the idea of liberation includes issues of personal and communal identity formation in a conflicted society. But here is the difficult confession: as noble as they may be, the ideas that we hold about freedom and liberation have not advanced the ultimate goals of peace, joy, and moral flourishing. Perhaps this is because we are relying on ubiquitous but thin definitions that are exemplified by personal choice.

If, as a people, we "ain't gone free," maybe it's because we have a very fuzzy and one-dimensional understanding of what freedom and liberation mean. The result is a vague sense of what we are seeking and what we believe. Perhaps we have not "gone free" because our definition of freedom is linked to uncertainty and its defining characteristics are gleaned from dominant cultural perspectives. We prefer black-and-white distinctions and definitions, but the idea of liberation and the quest for freedom seldom lend themselves to such a cut-and-dried analysis. In this book, I will be using the words *liberation* and *freedom* interchangeably with the caveat that a "trickster" element may be the source of apparent distinctions. To get the most of both ideas, it may be best to consider liberation and freedom as reflexive concepts that inform and challenge each other.

In the twenty-first century, the word "*liberation*" receives the same cultural disdain as the word "*liberal*." As a consequence, few people are willing to serve as role models. The result of such persistent and collective aversion is that liberators tend to work in the shadows. Finally, these noble concepts are often used to mask social and political shenanigans. As one of the major tenets of civil religion, the idea of freedom can serve as a conceptual shield to justify wars and imperialistic political agendas, for example, "they hate our freedom."

When this mantra was first recited after 9/11, heads nodded across the nation and affirmed this reason for the attacks. But what is it exactly that "they" hate? Does freedom have a particular configuration in the West that others hate? Is there a singular definition of freedom in America that applies to everyone? Does freedom mean the same for Donald Trump as it does for a homeless veteran? Did Lincoln "free" the slaves, and if so what exactly did that mean in the aftermath of the Civil War? The era that followed the purported act of liberation included Jim Crow, segregation, lynchings, and desperate poverty. Is this still freedom?

The questions persist. When we talk about freedom, are we all talking about the same thing? It is one thing to demand freedom but quite another to pinpoint the moment when it arrives. Freedom legislated and decreed is a poor substitute for the unfettered but interconnected state of being that seems to reflect the order of the universe. In twenty-first-century America, we hear more about freedom than we do about liberation. The assumption is that freedom can be dressed up in patriotic language in ways that support the domination systems, while liberation is the cry of every revolutionary and seldom behaves.

We know from history that liberation will eyeball injustice and challenge both liberators and dominators to view their own failings in the same mirror that they hold up to each other. Liberation is not just the freeing of persons and systems deemed to be "evil" or oppressive. It is also the opportunity to examine our own inner proclivities toward injustice. Oppression is an inner as well as an outer state of being and an equal opportunity activity.

This preference for "liberation" language is a recent occurrence. The elders who addressed injustice in years past struggled for freedom that was accessible to ordinary people. Every sharecropper and tenant farmer knew what it would mean to be free. There was less talk about liberation because the idea presumes effective agency. Although enslaved people have always exercised agency as they resisted domination systems in creative and sustained ways, they did not presume that those acts would ultimately liberate the community without the unlikely tools of music, art, and the intervention of a liberating God.

Music, art, and dream language helped to construct a new cosmology of liberation and to describe the indefatigable hope of the human spirit in the midst of degradation. In recent years, because of the excessive hagiography around Martin Luther King Jr.'s dream speech, folks are disdainful of dream language. Yet, during the Civil Rights movement, dream language spurred the downtrodden to break the taken-for-granted hold of oppression and to inspire marginalized people to imagine themselves free and whole.

Scholar Breyten Bretenbach agrees with this approach.

> And so we dream. There's the personal dream to come to terms with the inevitability of being finite. There's the communal one of justice and freedom upon which we hope to secure the survival of the group. And then there is the dimension of moral imagination. . . . I would postulate that we of this generation suffer from a massive failure of moral imagination. Instead of responsible freedom, we substituted self-enrichment and entitlement linked to cowardice, bad faith, the corruption of dependence, and the glorification of impotence or of posturing expressed as political correctness, whereby our languages were gutted of texture and color.[6]

This generation is being called upon to recover moral imagination and the rich nuances of storytelling and dream language in preparation for a lasting liberation. After many false starts, it is clear that we will not find our way to this destination through the courts or through our protests. Instead, we must dream, sing, and imagine the scope of this social experiment beyond the boundaries of mundane oppressions. Because this is no easy task, the arts are critical to liberation. We saw this in the mid-twentieth century when music became one of the most important components of the Civil Rights movement.

Music incites our moral imagination. We need only hear a tune to imagine where we were or who we were with and what was going on at the time. It was the music in South Africa and the rhythm of resolute feet dancing the *toyi toyi* that unraveled apartheid. The rhythm of like minds and common stories bound by the music of revolution can topple governments. At the height of the Civil Rights movement in the United States, Nina Simone sang Dr. Billy Taylor's composition "I Wish I Knew How It Would Feel to Be Free."[7]

The song was never as popular as the Civil Rights movement anthem "We Shall Overcome," but its lyrics spoke volumes about the enigma that liberation poses for the collective imagination of those seeking to be free. By contrast, "We Shall Overcome" was first sung as a gospel song and later became the rallying tune of labor and civil rights activists.[8] As an anthem of intent, it attested to "the evidence of things not seen" despite seemingly insurmountable odds.

"We Shall Overcome" was exactly the song needed to inspire the Civil Rights movement and the emergence of black liberation theology.

It was a musical first step toward wholeness and a communal departure from familiar abuse. But that was then and this is now. Today, it remains a popular expression of public commitment to justice and equality. However, the sentimentality of that tune grates against the deeply sedimented reality of racism, and its status as a much-loved anthem has eroded into a sentimental musical group hug. It is now apparent that the "someday" language in the song obliquely blesses hopeful uncertainty as the spiritual vehicle for the liberation of the oppressed. I understand the intent to posit the "someday" deadline as an already/not yet goal predicated on the faith of long-suffering people. But it is another matter indeed to assign a permanent status to this song as the signifier of the liberation of black folks.

Both freedom songs articulate important and relevant aspects of liberation in imaginative ways, and both note the elusiveness of freedom. We don't know when it will arrive, and when it finally comes, we may not even recognize it because we don't always know what it means to be free. Freedom and liberation are intertwined strands that represent the human attempt to align our limited and temporary social experiments with the justice that seems to be endemic to the universe.

Liberation and Communal Solidarity

The greatest hope of a domination system that pretends invisibility and seems to have no origin and no parentage is that each person, each community, will seek a freedom from bondage that benefits only one, some, or a few. Such liberations offer the sustainability of cut flowers, beautiful for only a short time. How can liberation set down roots if it is treated as the property of one group and not another? At times, the black liberation movement has resisted alliances with other justice initiatives on the basis that race is the most important oppression. In the twenty-first century, we are reluctant to talk about a black community that is fractured on the anvil of sexuality and gender. We are liberal about issues of race while we willingly align ourselves with the forces of oppression when issues of gender and sexuality are on the table. In response, scholars exposed the structurally intertwined reality of oppression. Unfortunately, although this work offers rich intellectual engagement, it seldom vaults the church/academy divide. And so, it is possible to have mega-churches in the black community who battle gentrification and economic injustice while marching against the rights of the gay, lesbian, bisexual, transgender (GLBT) community.

We can agree in principle that all people should be free, but when choices must be made in favor of a neighbor's needs, we forget that varied forms of oppression are interlocking pieces of a puzzle. We tend to prioritize the injustices that have the greatest impact on our own lives, but there is more at stake. Issues of gender, race, class, and sexuality may ignite different passions, but when viewed as part of a whole, it becomes apparent that effective resistance requires unity. Lasting liberations require allies in solidarity against oppression in any form.

Solidarity favors the well-being of the whole over personal preference and theological misreadings and myopia. It also requires a cognizance about cosmological origins and shared destiny. When solidarity is rejected today, it is usually on the basis of some moral or theological objection. Christians will read the Bible as a divine mandate to reject the GLBT liberation agenda but will choose a looser and more politically correct reading of passages in the Hebrew Bible that separate and discriminate against the dis/abled. Solidarity lays the groundwork for liberation because it is dangerous to the systems of oppression.

Finally, solidarity is a choice for fluidity and dialogue. It allows for critique, disagreement, and debate. New alliances with other freedom seekers is not just an external boundary crossing; it also shatters the illusion of intragroup unity and brings support to the unacknowledged difference that is always a part of every community.

I was beginning to understand that the narrative form of this book will support the elders' advocacy for the return of moral imagination. From what I understand, the ancestors also intend to situate notions of liberation in generational, cosmological, and wisdom-oriented contexts so that future generations will know how it feels to be free. But I still needed some answers, and it wasn't long before I became aware of Sarah's presence.

Writer I have been waiting for you to show up, I have been thinking about the philosophy of liberation and the liberation/freedom distinction.

Sarah The meeting is being called because it is clear that the generations who followed the civil rights era have only a peripheral understanding of liberation and even less moral imagination. The end of Jim Crow and segregation heralded the end of one form of oppression but did not end racism once and for all and did not describe the shape or content of liberation. Although civil rights strategies helped black citizens to obtain rights that should have been granted under the Constitution, those strategies were limited to the social and political realities of the mid-twentieth century.

What was needed was an understanding of freedom as an event that would evolve and change to meet the needs of future generations. What is needed is an idea of liberation that is linked to divine and cosmological realities with a specific blueprint for social action that includes but is not limited to protests and marches. Liberation in a cosmological context includes considerations of interconnections, realignments of identity, fluid alliances, and an interactive reconciliation process. It is not limited to personal desire. But that's not what I came here to tell you. This is the beginning of the book, so can we lighten up a bit?

The interviews have not even begun yet and you are mulling over a few centuries' worth of freedom movements, dissing the civil rights anthem, and generally putting all of us to sleep. I know that you want answers, but you won't get any if you bore us to death. The definitions will emerge during the conversations. And for your information, while you are waxing eloquent about what you think about liberation and freedom, the elders are waiting.

Writer All right, but can we resolve a few things before the interviews begin? Let's use the title of this chapter to guide our conversation. It won't take long. I'm as eager as you are to begin the interviews. I just need to understand why we have not "gone free" in any permanent sense, and I also need a better understanding of how liberation connects to cosmology. Also, it seems that there are cycles of liberation instead of one final and complete attainment. What is keeping us from embodying the freedom that seems to be endemic to the universe?

Sarah All right, that seems fair, but just for a few more minutes. I may have to bring someone else in to talk cosmology. That's beyond my expertise.

Writer This whole project is designed to discuss the liberation project and its future. It will be interesting to hear what the elders have to say to this generation, because the idea of liberation has a different meaning now than it did one hundred years ago. Our generation no longer flees cotton plantations; instead, some of us are uncertain as to whether we want to be free or whether we want to join the ranks of the oppressors and share the spoils.

Sarah That decision has to be made. Freedom is a choice. The God of the Universe models freedom then beckons all of us toward the human version of a liberated life.

It's not that we don't understand the options, because messengers bear witness and invite us to view the wonders of the universe, and there are clues about freedom preserved in every holy book on earth. And yet human beings struggle to attain it. We sing about freedom and fight for it, and yet it alludes us. One can almost hear God saying, "Ain't ya free yet, why ain't ya gone free?"

In recent years, liberation theology offered an open invitation to continue the exodus from the comfort of subservience and nonbeing toward self-actualization and communal empowerment. There is nothing practical about the act of liberation. The desire for freedom necessitates a leap toward transcendence and an understanding that being free is an unfolding and processive event.

Writer Processive? Are we talking about process theology now?

Sarah Yes and no. In the scheme of things, liberation theologies are relatively new, and because of this reality, Rufus Burrow notes that "liberation methodology remains in a developing or processive mode."[9] The word *processive* has more to do with movement, openness, and the ongoing nature of liberation paradigms. What I am referring to as a processive approach to theology can be traced in some measure back to the enlightened parentage of Whitehead and Hartshorne, systematic theology, creative imagination, and the lure of a God who is "in it with us." The pastiche that results is exactly the fertile repository needed to nurture the idea of freedom.

The idea of freedom that inspires black liberation theologies also emerges from the alchemy of critical engagement with Western scholarship, black church traditions, ancestral inclinations as dream legacies, slave survival genius as creative adaptation, and the proverbs of the "big mamas"s. These resources are very open and processive.

Although freedom is not defined or confined by race, color, gender, class, or sexuality, each unique manifestation advances the wisdom that we have about our destiny as liberated souls. There are underlying similarities between racial/ethnic and dominant cultural appropriations, but black, Asian, feminist/womanist, and Hispanic theologies are not just a deeply tanned or exotic version of white theology. They are processive theological inquiries into the very nature of our being.

Writer Can I take a turn at describing this free-flowing but deeply interconnected process?

Sarah Why not?

Writer Well, from what you've just said, processive approaches to liberation would evince openness, communal resolve, and energy as a modality of social change; the interplay of temporality, liminality, and ultimacy in the life space; a recognition of the intelligence and intelligibility of the cosmos; our cyclical human emergence therein; and a respect for the arts as a source of potential and creative engagement and not just as devices for temporary amusements. These things seem doable, and yet freedom remains illusive.

Sarah Yes. That's why the God identified and embraced by black, Hispanic, and Asian theologies is as surprised as you are that bondage is easier to accept than freedom.

Writer I want to return to this idea of "processiveness," but right now I'm extremely curious about your allusions to God. There is a vastness inherent in your references that make our church depictions of God pale by comparison. When you refer to God, I hope that you are not thinking of an "old man in the sky" because that personification is troubling to me.

Sarah It is troubling to me too, especially on my side of reality. Although I have no words to explain the reality of God, I can say that God is the wisdom of every lifetime, a deep plunge into a clear pool, the sinew and muscle of ethical responsibility, a community of goodness, but always more. Descriptions reach out as far as they can toward the God of the universe, and then, like a rubber band stretched too far, they snap back and we are left with the silence of mystery and awe.

Writer But isn't silence better given our penchant for misconstruing everything?

Sarah The antidote is immersion not aversion. It is the willingness to fully immerse ourselves in the community where a lasting and transcendent liberation will emerge. We are creatures who come from the unknown and reenter that same realm upon death. What can be lost if we spend the time sandwiched between these question marks as advocates of a form of justice that is grounded in the mysteries of the cosmos and the needs of the community?

> The God of the biblical witness is not Zeus who seeks to compete with human beings by using mythological or ontological chains to bind them to an unchangeable spot in the universe. The God of Scripture calls humanity to freedom and to a covenant partnership that is redemptive and liberating.[10]

Sarah We are nudged by divine intention toward a model of freedom that is situated not only in the body but also in the mind/consciousness. Renewal occurs as we dive into the depths of God and expand our sense of reality to include a rapidly expanding universe. The move toward processive liberation is communal, cosmological, and contemplative. But to achieve this expanded perspective, this generation must shift its focus from the mundane. As Barbara Brown Taylor offers:

> If nothing else, reading a little cosmology now and then is a good corrective for those of us who speak too easily of God. If we really believe the one whom we worship is the creator of heaven and earth, then where do we get the nerve to offer tidy explanations about exactly who that one is and exactly how that one acts?[11]

Writer Well, that's all well and good, but I have found that people of color are cautious about moving from deterministic to cosmological freedoms because they know that while the cosmological turn is being proposed, dominant systems are using up most of the world's assets, exploiting the labor and natural resources of most of the first nations of the world.

Sarah I understand that. Oppressed people should be suspicious when dominant orders tell them to take their eyes off of the specific elements of their exploitation in favor of a view of the cosmos reflecting on itself. In the meantime, corporate interests are continuing their acquisitions.

Writer While they are stargazing, more factories to make Nike shoes are opening in Thailand, someone has a hand in your pocket, and Western magicians are hiding your oil, and pharmaceutical plants, up their collective sleeves.

Sarah All true! But our task is not to protect assets; it is to find wholeness. Humankind's story of liberation must connect to the larger story of a God who is free and a universe whose moral arc bends toward justice. The stories that you will be hearing from the elders are not told in isolation. The universe also has a story. It is a story of creativity, of careful balances threaded between life and extinction, an aura of purposefulness that arises from our own observations of the thin margins decided in favor of life on earth.

> Everything that exists in the universe came from a common origin. The material of your body and the material of my body are intrinsically related because they emerged from and are caught up in a single energetic event. . . . This universe is a single multiform energetic unfolding of matter, mind, intelligence, and life. And all of this is new. . . . We are the first generation to live with an empirical view of the origin of the universe. We are the first humans to look into the night sky and see the birth of the stars, the birth of the galaxies, the birth of the cosmos as a whole. Our future as a species will be forged within this new story of the world.[12]

Writer So the ancestors are coming to remind us that our story is also the story of the cosmos.

Sarah Yes. They knew this well before their transition. During their lifetimes, they never really converted to a way of knowing that pinpointed the source of life and liberation in the mundane. They will suggest that liberation was configured in the twentieth century as a static goal that should be supplanted by a more dynamic idea. The ancestors will reintroduce this generation to a type of liberation that is cosmologically grounded, inclusive of other liberation movements, reconciling and reflexive. Through the stories and conversations, we will be taught to reclaim the past as well as the present and to stitch our differences into shared initiatives through the synergy of dialogue.

To say that liberation is reconciling is to note that it does not exalt over defeated enemies but instead invites them into the hard-won victory as participants. This type of reconciliation does not humiliate victims through the cheap and easy forgiveness that is offered to an opponent so that the conversation will end. Such moments are never reconciling; instead, they seal the sight of the infected wound under the theory that if we don't see it, and if we don't talk about it, perhaps it will go away. Inevitably, the infected community either dies or erupts, but the hope of reunion ends with the false reconciliation.

During the liberation movements of the mid-twentieth century, the consciousness of the nation was raised and laws were passed, but the complexity of liberation in other contexts was not considered. Inclusivity would have required concern for other oppressed communities, which includes the earth community and its nonhuman members; reflexivity would have kept historical amnesia at bay by linking the past and the future through dialogue; and reconciling initiatives would have aligned personal interests with the needs of global neighbors. Real reconciliation is grounded in fairness and truth telling.

Writer I understand what you're saying. Static formulations of liberation configure the opportunities for transformation within specific communities and nowhere else. Such limited political and social perspectives make liberation as a permanent ground of being impossible. A processive view presumes that the liberation of the human community is connected to the freedom of all beings and all living things.

Sarah Yes, freedom feels personal because it conforms to the shape of the need in the seekers, but its reach exceeds the individual. To have settled for individual rights when freedom was available was a mistake. There was a moment in time when the domination systems paused and seemed willing to shift slightly toward justice.

Writer That window of opportunity has closed. Protests no longer work; the political system is captive to gerrymandering and the media message. Even worse, the most heinous of government actions can through talking points be given a positive spin, and each group that seeks freedom wants to believe that its movement is the pinnacle of all justice pursuits—hence, the resentment toward emerging justice seekers.

Whether we like it or, liberation connects us to everything, even the people with whom we disagree. We have to be concerned about animal rights, the rapidly expanding universe, the plight of Mexican immigrants, GLBT liberation movements, and wars on terror not because all of those causes personally benefit us but because we are deeply connected.

Writer What you seem to be saying is that the pursuit of freedom has the best chance of success when it emanates from a community willing to see itself as others see it, and willing to hear cries of dissent even while anthems of empire are trumpeted from sea to shining sea. The desire to be free is not confined to one species or one community.

Sarah Grounding liberation in cosmology would require that you evaluate human actions in the world using a more comprehensive perspective. Every war would have to be measured not just against the freedom interests of the offended nation but also against the freedom of innocents not to be bombed or tortured into oblivion as collateral damage. Because the idea of liberation is a multidimensional event, it may not always manifest in the same ways, but it usually leads to that state of being that we refer to as freedom. And, freedom is like a hologram, because it encodes the larger philosophical ideal in its smaller and more specific ideas.

What people refuse to entertain is that various ideas of freedom can coexist and even teach one another if the initiating premise is morally grounded. A great example can be found in our increasing awareness that the West consumes more goods and services than any other nation on earth. If this pattern of consumption is based on an understanding of freedom that is ethnocentric and greed oriented, then it will grate against the need of emerging nations for clean water and sustainable living conditions.

Perhaps what the West calls freedom is not freedom at all, because freedom, like love, is not self-serving or obsessed with self-interest. In the West, we use the word *freedom* in the same way that one can name a pig a rose, but when that pig begins to wallow in the mud, you will know that you won't be making any corsages with it. A state of sociopolitical existence that rapaciously gobbles up all other forms of social organization is not freedom no matter how many freedom songs are sung.

Writer Isn't it our task as part of the cosmological move toward liberation to situate ourselves as citizens of the cosmos instead of as patriots of one nation or another? To do otherwise limits our view of liberation.

Sarah The story of the universe is layered with the said and unsaid, with what we know and that which cannot be known or deciphered but that on occasion, through a transcendence of mind and spirit, can be discerned. To ignore the story of the universe is to limit the potential for freedom. The idea of liberation that we have been talking about lies in the acknowledgement that we are more than we have chosen to believe about ourselves. "More" is always more difficult than "less." "More" requires something extraordinary from the human family. It is the requirement that you remain grounded in everyday reality while leaping toward the future. It is the business of personal and communal focus and concern and at the same time the enlargement of identity and purpose to include the bigger picture.

Writer This is getting complicated. Can we make a few more cosmological connections and then move on?

Einstein Did you ring?

Writer I'm stunned. Sarah, where are you? If this is actually Einstein, I can't have this conversation. How can I talk with someone of his scientific stature?

Sarah You'd better get used to the red carpet. A bunch of famous folks are waiting to talk with you. Put on your game face and get ready. [Turning to the wild-haired man] Hi "E," what's going on?

Einstein I couldn't be better. You can't imagine what it's like to be in the midst of a reality that you saw only mathematically during your lifetime. I know that I'm not part of this conversation, but moral imagination invites unlikely suspects into the mix. I just thought that I'd drop some science before I get back to work on the theory of everything.

Writer Why are you still working on theories? Don't you know that you are on the spirit side?

Einstein Where do you think that my ideas came from when I was on the earth side? In the spirit realm, imagination and creativity are nurtured and passed to the other side through dreams and flashes of insight and intuition. My ideas came like that, whole and shared. I can do no less for the next generation. Some young physicist or cosmologist is puzzling right now over a problem that seems to have no solution. And if the purportedly rational path of inquiry is the only tool in her or his toolbox, the breakthrough may never come.

There is a mind of inquiry (which some call consciousness) that all of us can tap into if we release our presumptions about the limitations of our own brains. If I had time, I'd do some work on this theory. Roger Nelson has his work cut out for him with the global consciousness project because he is tapping into ancient ideas that are still beyond the pale of postmodern science. Science is as much about agreement as anything else. Once things are named and categorized, and then published and discussed at prestigious conferences, they are hard to displace. The maxims that we come up with as a result of this process seldom do more than set forth a series of things that we agree on, but that speaks more about what we believe about the universe than it does about what the universe is really like.

At age sixteen, I saw myself chasing a beam of light, and the rest is history. Although I was unaware of it at the time, this flash of insight would require a completely new thought system. To unveil the theories of general and special relativity, I would have to chart a new path. I could neither tinker with the emission theory of light nor retrofit the electrodynamic theory. Everything was up for grabs, including our settled theories regarding space and time.[13]

You are looking for liberation using the old language and the old formulas that launched the movements in the 1960s. You will have to chart a new path. You cannot retrofit the march or sit-in to be viable options in the twenty-first century. You can't recycle strategies that were successful against naive oppressors while in the midst of a technological revolution.

What you are seeking is the norm, not the anomaly. Liberation is integral to the cosmos, and so are our religious inclinations, for that matter. It's hard to see the connections because cosmology is offered as an elite science completely alien to ordinary people and because religion is practiced in a way that is intended to make people comfortable.

In each case, the inclination toward elitism and comfort hampers the fluidity and dynamism that both were intended to have. I would not have said that a few years ago. I was a great proponent of organized and deterministic systems until the universe began to reveal itself to me as interconnected, intelligible, and free.

Our bodies are part of the articulated stardust of the universe. We live in a planetary ecosystem with its own environments and destiny. Politics notwithstanding, we have the potential to change everything through our reconnection to that power source.

The first liberations should be from the belief that we are limited, small, and inconsequential. As John Polkinghorne likes to say, the intelligence of the universe can be known only through the reflective gaze of humankind. Well, he doesn't exactly put it that way, but the thought is the same. John Archibald Wheeler makes similar claims:

> [Wheeler, a] student of Niels Bohr, teacher of Richard Feynman
> . . . has dared to asked some of the Biggest questions imaginable

about the universe. . . . [He] is reputed to have said: The "U" in universe stands for "you." You are the Universe, looking in the mirror. When you observe the Universe, you are observing U-rself.[14]

Writer I understand what you are saying, but at the same time I don't.

Einstein That's because there is an internal struggle regarding truth that occurs even for those who don't know the intricacies of scientific discourse. I am offering explanations that don't comport with the stories that you were told about the world in second grade. You are asking for simple explanations, but simplicity is found in intricacy and vice versa. Cosmology and liberation are connected because they speak to the underlying nature of reality.

Unless you have a few centuries to spare (and I actually do at this point), all I can do is offer you a telescope on a dark clear night. Who are you when you look through the lens? What issues matter when you view the Milky Way? Everybody is down on teens and their gang activities, but aren't the adults doing the same? They just use the earth as their "hood." Look within and gaze beyond if you want to find the clues to freedom. The quest for liberation may require that you relinquish your deep-rooted attachments to material things; it may mean that you embrace the mystery of darkness in the universe in the same way that you seek the power of light. But the hardest requirement of all may be to release carefully constructed and protected self-images to imagine yourself as God imagines you and as you will be when your human body no longer confines you.

Writer But what did you mean when you said that we are connected?

Einstein There is no time for a detailed explanation. I hear that you have a meeting to attend. But if I were to sum it up quickly, I would say that the scientific concept of nonlocality gives us clues as to the interconnected nature of reality. I can define the concept for you, but that definition must remain fluid, while the universe slowly and reluctantly reveals its secrets.

> The essential new quality implied by the quantum theory is non-locality; i.e. that a system cannot be analyzed into parts whose basic properties do not depend on the . . . whole system. . . . This leads to the radically new notion of unbroken wholeness of the entire universe.[15]

This theory of wholeness is very inconvenient for those who have staked their lives on individualism and self-actualization, but a blessing to those who seek a lasting and shared liberation. If quantum theory has a cultural impact, it may mean that we affect one another without respect to distance or difference and that rumors of our independence are greatly exaggerated. Think about it. I've got to run.

Writer I understand that we inhabit a life space that includes quantum elements that seem to appear freely when and where they want. I'm also beginning to understand that we are citizens of the universe, not just of a nation state.

At this point, I turned to ask another question about what it meant to be stardust, but Einstein was gone.

Writer All right, perhaps liberation is not the impossible dream. There is an order to the universe that looks like freedom, but this is not the freedom of our dreams. Instead, it is the invitation to journey toward newness.

It was important that the first steps toward liberation included rulings handed down through federal courts, but the law was never intended to be the last word. The liberation that the prophets speak of offers a beginning for people of faith that is processive and ongoing. I am ready to consider a more expansive and cosmological liberation, and I'm ready for the conversations to start. Hey, Sarah, where can I find a telescope?

Scattered throughout the apparently hostile masses which are fighting each other, there are elements everywhere which are only waiting for a shock in order to re-orient themselves and unite.
—Teilhard de Chardin

3. Law and Liberation: A Love Story?

If there were love of neighbor, there would be no selfishness, none of such cruel inequalities in society, no abductions, no crimes.
—Bishop Oscar Romero

Barbara Jordan and Thurgood Marshall were legal insiders and sociocultural outsiders. They grew up in African American communities shrouded from public view by segregation. Both engaged in the pursuit of liberty seeded in the laws, Constitution, and judiciary of the United States. During the Civil Rights movement and in the decades that followed, they considered their legal work to be as important as the work of activists and protestors. Both believed that social change depends on the spirit of liberation encoded in the laws of the nation. In this chapter, Jordan and Marshall return to assess the legacy of civil rights legislation and the future of law as an ally of liberation.

The Elders

Barbara Jordan, 1936–1996

There is an interconnectedness between each of us which cannot be abolished, not withstanding our personal wishes and desires.[1]

Barbara Jordan was born February 21, 1936, in Houston, Texas. Her life as an African American woman in a segregated society unfolds in unpredictable ways. She is remembered because of one moment in time (July 25, 1974) when she called the president of the United States into accountability without ever mentioning his name. In her "Opening Statement to the House Judiciary Committee Proceedings," she reminded Richard Nixon and a listening nation that the United States was built on principles of moral integrity and a presumption of the good faith and good character of its leaders. Her exhortation in favor of the impeachment of any leader who violated the Constitution accomplished what a free press, a special prosecutor, and top-level resignations had not. Soon thereafter, an out-of-control president finally resigned. Today, the situation seems more like a comedy of errors than a national moral crisis, but we are viewing the events retrospectively.

Today, we are used to seeing people of every ethnic origin in public life. This was not the case when Jordan became the national spokesperson for ethics and integrity. At the time, African American women were not expected participants in public debate. It was not unusual to read a newspaper article that described her as an Aunt Jemima who sounded like God. Notwithstanding the public insults and commendations, Jordan became the face of a new articulate, political, and hopeful generation.

But there was more to her life and her accomplishments than the shock of that televised moment. She delivered two keynote speeches at Democratic national conventions, chaired the U.S. Commission on Immigration Reform, taught ethics at the Lyndon Johnson School of Public Affairs at the University of Texas-Austin, and received the Presidential Medal of Freedom.

There is no question that Jordan was the quintessential insider politician, but she believed in a God of liberation and in the potential for freedom inscribed in the Constitution. Her view of liberation exceeded release from captivity; rather, it emanated from the ideals set forth by slaveholding nation-founders who described an egalitarian national community that they could not enact or exemplify. Jordan believed that the constitutional precepts of equality and freedom were attainable and worthy of our mutual striving. Through her ability with words, she ignited a public conversation about liberation that continues to this day.

Thurgood Marshall, 1908-1993

Liberty cannot bloom amid hate.[2]

Those of us who came of age during the mid-twentieth century remember Thurgood Marshall as the crusty and irascible jurist who seemed to be more like an Old Testament prophet than a Supreme Court justice. He ended his years on the bench resisting the death penalty in all cases and speaking truth to whoever would listen while he wondered about the staying power of the struggle for liberation. It was heartbreaking to see Marshall replaced on the Supreme Court by a man whose politics and commitments were as opposite to his as possible.

Thurgood Marshall was born in Baltimore, Maryland, on July 2, 1908. Although it seems that slavery is now so far from our daily lives as to be irrelevant, Marshall was the grandson of a slave, the son of a railroad porter and a schoolteacher. Educated at Lincoln and Howard University, he was mentored by Charles Hamilton Houston. It was this relationship that shaped his legal career and eventually situated him in the center of the battle for civil rights.

Marshall's life is punctuated with "firsts"—first black Supreme Court justice, first black lawyer to win a case before the Supreme Court, first in his law school class—but his life speaks of more than personal accomplishments. Marshall was committed to an ongoing struggle for the liberation of the oppressed, whether they were being denied entrance to a state university or relegated to death row because their poverty would not allow proper representation. Marshall served as president of the National Association for the Advancement of Colored People (NAACP) from 1940 to 1961. These are the critical years of struggle for inclusion and respect. He also participated in the drafting of constitutions for nascent African countries, and he argued *Brown v. Board of Education*. Like Barbara Jordan, he believed in the Constitution but challenged the applications and interpretations that negatively impacted the lives of minorities.

Marshall's view of liberation was that it is attainable if people of conscience engage the powers of oppression. He viewed his personal opportunities and accomplishments as tools for the struggle on behalf of those who had no representation. Unlike Jordan, he did not see the potential for liberation in the words of the Constitution of the United States. Marshall believed that the Constitution was a good effort but flawed in its ability to describe and set a pattern for a free and liberated society. The guiding principle of his legal career and public life was that you talk the talk, *and* walk the walk. He would be disappointed in his replacement but would be encouraged by the increasing diversification of the American polis. Diversity, after all, is the template for freedom.

The Conversation

This chapter is about the relationship between law and liberation. Certainly, law and liberation are entangled in a complex relationship. As in any relationship, there are good days and bad. Sometimes they murmur their love and support for each other; at other times; they square off, accusing each other of cheating. During the 1960s, there was great hope that they would live happily ever after, but the road has been rocky and the estrangement seems permanent. Law says to liberation, "Why must you be so brash and loud?" and liberation says to law, "You have a commitment phobia. If you did more than pontificate, we might already have achieved the justice that we both want."

Each makes a case and yearns for the embrace of the other. Law admits that while Jim Crow juridically constricted the marginalized and forced them to contend for their freedom, it was the legal codification of freedom's ideal in the United States Constitution that inspired the struggle for equality. Liberation proudly points to the nonviolent resistance of the Civil Rights movement as one of its finest hours but admits that marches do wear thin in both inspiration and effectiveness.

It is this relationship between law and liberation that is really at stake in today's conversation. The overarching question that will guide this discussion is whether the relationship between liberation and law is really a love story. There is something at stake that is bigger than case law and precedent. When Thurgood Marshall was fighting the death penalty by dissenting every time the majority upheld the right to execute a person, it wasn't to save on the costs of lethal injections. It was to save the very heart of a national ideal. Every time an impoverished, poorly underrepresented inmate was executed, the ideal of equality under the law and the spirit of mutuality and communal care suffered catastrophically. Marshall understood that the state was not just killing individuals for one crime or another; it was killing the spirit of the nation in our names.

> Laws are generally found to be nets of such a texture. As the little creep through, the great break through, and the middle-sized are alone entangled in.[3]

Jordan's idea of liberation emerged from her love of the Constitution, an odd affection, to be certain, but heartfelt nonetheless. She saw liberation peeking from constitutional codifications even when the ideal did not match the reality of life in the United States. Today, I would be talking with Jordan and Marshall about liberation and law in purportedly civil societies and in an expansive universe. Both elders openly opposed legal and social infringements upon freedom, and both believed that liberation required action in the courts as well as the streets. I was musing about their amazing contributions when the interview began without fanfare or announcement. It was just as well that I didn't know what to expect; the surprises that were to come enhanced the experience.

If it were not for the trademark voice, I would not have recognized Barbara Jordan. She didn't appear to be the woman that I remembered: solid, impressive, and no nonsense. The "no nonsense" part was still intact, but the rest was less familiar. Jordan looked like a triathlon athlete in the winner's circle after a race. She was lithe, playful, and strong. When Jordan was growing up, she recalled that her embodiment was deemed to be a detriment.

> The whole system . . . was saying to us that you achieved more, you went further, you had a better chance . . . if you were not black-black with kinky hair. Black was bad and you didn't want to be black, and so the message was that it was too bad that you were so unfortunate that your skin was totally black and there was no light anywhere.[4]

Despite the pejorative characterization, she never accepted the labels that society tried to assign to her. She was a strong and focused presence in the world. Now her spiritual embodiment reflected that reality. This was my first interview. I would soon learn that these luminaries seemed to have taken on a spiritual "shape." The transition from earthbound to ancestral lives changed their "bodies" but amplified their spirits and personalities.

Everyone that I interviewed seemed to be turned inside out. I could see what was on the inside. And this spiritual shape superseded my memory of their physical shape. It made sense that Jordan looked more like the runner Althea Gibson than herself. Now, she embodied the strength and power of her mind and spirit. I make no presumptions about the meaning of all of this. For all I know, they may have looked exactly as they did in life. The change may have been in my perception. I don't know; I can only report what I perceive.

Jordan I have a feeling that you want me to comment on the ruling of the Supreme Court in *Bush v. Gore*.[5] It is a natural response. I became a household name when I challenged Nixon during Watergate. I must say that I am intrigued by the first elections of the twenty-first century.

She certainly knew how to get to the point. But before I could say anything, Thurgood walked in. He gave the impression of dapper sophistication and charm.

Marshall Well, I came in at just the right time. Don't get me started on the subject of *Bush v. Gore*.

Writer [I wasn't certain that I wanted to open that can of worms.] Look, because this is the first interview, I don't want to begin with a partisan political attack on one side or the other. Besides, the Bush administration is ending, and a woman and an African American are making history as candidates for the presidency. A new day may be coming. So, with all due respect, if you want this message to get out, I'm going to suggest that you entice readers from all camps.

Thurgood laughed and reminded me that the elders had specific responsibilities during these conversations.

Marshall This is not a partisan topic for the conversation; it is history making. Your task, as I understand it, is to write what you hear, and this is what you're going to hear from me. Since when does the Court decide elections?

Writer [I let the moment pass without answering.]

Jordan Was it appropriate to substitute the judgment of the people for the judgment of the courts, and why would the highest court in the land enter into this partisan fray?

Marshall I like what David A. Strauss says in an essay entitled *"Bush v. Gore*: What Were They Thinking?"

> Judges, like everyone else, sometimes act on instinct. That is inevitable and, as I said at the outset, often unobjectionable, within limits. But a close election is sure to inflame partisan passions and skew judgment in a partisan direction. That makes it all the more important for judges to hesitate, to question their own motives, and to make sure that their judgments have a solid basis in the law before they act. That may be a lot to expect.

But if there is any court of which we should expect it, it is the United States Supreme Court. In *Bush v. Gore*, a majority of the Court, prompted by a general and unjustified sense that something needed to be done, plunged in, splintered along ideological lines, and played a prominent role in deciding the election. This was not a triumph for the rule of law.[6]

I couldn't have said it better myself. The nation has to believe that the court has a moral center that cannot be influenced by anyone or anything. It might have been better to let the impasse continue until all of the facts were in. I have no dog in this fight, but I do have a sense that Supreme Court decisions are breadcrumbs dropped in the midst of a social forest that point the way out of the thicket.

Jordan I just have to say this—the politicization of the Supreme Court is shocking. There were very few things in Washington, D.C., that did not bear the imprint of politics, but the overt use of the court to determine an election leaves me without words.

Marshall I don't believe that for one moment. I have never seen you at a loss for words. Barbara, your words have words. But I do agree that a judicially selected president fundamentally changes the nation. I am not alleging wrongdoing, because legally everything seems to have been handled according to the books. But something is missing.

It is the shared understanding that some of our choices, although perfectly legal, may contravene the spirit of the Constitution and the laws. On occasion, we forget this balancing factor. It is during such times that the press and the people are supposed to help put things back into perspective.

During his lifetime, Marshall's cries for justice were sometimes strident, as he battled against the death penalty and the erosion of civil rights right up until the end. Now his remarks were tempered with a philosophical turn of phrase that Jordan found downright funny. In between serious statements, he was chiding her with inside legal jokes and she was enjoying every moment. Oh, how she could laugh. When the joking stopped, Marshall became pensive. I thought that I'd better get a question in before the bantering took off again.

Marshall It may seem odd to you that we can discuss something as serious as *Bush v. Gore* and still tell jokes on the side, but sometimes you have to laugh to keep from crying.

Jordan, who was known to love a good sing along, began to hum "Cry Me a River" under her breath, and they both laughed.

Jordan Seriously, though, there seems to be a silence in the land. If you want to find and enhance a liberation that is more than the by-product of political maneuverings, then you are talking about making some noise. The business of public dialogue requires the risk of conflict and disagreement but promises an airing of issues that affect the common good. Weaving tapestries of public debate and good citizenship is not easy business.

Marshall I want to think about this a bit more, so can we come back to that subject? Right now, I have a bone to pick with you, Barbara.

Writer Me? [I wondered what I could have done wrong so early in the game.]

Marshall [shaking his head] There are too many women named Barbara in this room. I'm talking to the gentlewoman from Texas. I hear that you kept a ratty copy of the Constitution in your pocket until the day that you died. What good is paper when the reality is so far from the written ideals? Toward the end, I wasn't certain that we were doing any good in our efforts to ensure the liberty of future generations. If we are honest about it, we really couldn't determine what that liberty should look like and who should reap its benefits.

Jordan But isn't that just the point, Justice Marshall? [Her voice was formal and commanding, as if she were standing before him arguing a case.]

Marshall Don't take that tone with me, young lady. We're just a couple of folks talking. But seriously, did you really love the Constitution?

Jordan Love is a funny thing, Thurgood. [Jordan seemed to be remembering something from another time and place.] I loved the Constitution because it was written in a way that reminded me of biblical texts.

Marshall Oh, come on, Barbara. [Thurgood was getting riled.] I can't let you get away with that.

Jordan [raising her hand and speaking quietly] Hear me out. The writers of the Constitution were slaveholders but they wrote of a liberty and freedom that was on the tips of their pens but not yet in their hearts. The biblical stories speak of moral values that we have yet to fulfill. Some documents seem to be spiritually inspired, imbued with good intentions for humankind even though the counter testimony of lives lived in opposition to the document can forestall or mute its prophetic message. The

Constitution bears prophetic messages that I carried with me as a reminder of what the God of the universe wanted for me and others like me.

Marshall I still think that the language of love should be reserved for relationships that involve reciprocity and mutual care. I would argue that neither the law nor the Constitution comes close to the love of God or to God's hope for our ultimate liberation. My God is more. . . OK, let me use the language of this project . . . my God is more "cosmological" than that.

In essence, the laws of the United States only encode the desires of its people. During our lifetime and the present time, the people seem to be perfectly comfortable with tiered citizenship and unequal applications of the laws for the benefit of the wealthy and to the detriment of the poor.

Everything in us rails against what I just said because we don't want to believe that we have encoded unfairness into the system, but we have. If you were to tell me that you love the Constitution and the body of laws because they are better than the laws of some countries or because sometimes we do the right thing, I'd say fine.

It was time for me to intercede. I assured them that I was fascinated by the reminiscences and discussion but reminded them that I needed to hear from them about the liberation project and the law in America. I didn't know how I was going to get to the cosmological with Jordan and Marshall, but perhaps the connections could be made.

Writer Let me ask you the big question first. You know where we are. what do you think of the state of justice in the United States today. Representative Jordan, could you address the interface between public discourse and public theology? Oh yes, and Justice Marshall, it is clear that we cannot go on with business as usual? What does the future hold?

Marshall [a glint of sheer delight in his eyes] Wrong office, my dear. I'm not God and don't want to be. The future is in your hands. You are creating it every day. Surely you've heard about the cycles of sowing and reaping.

Writer Yes, but, . . . from where you sit you must have a sense of the big picture. You know how the story ends.

Marshall So do you. Don't Christians know how the story ends?

Jordan Now wait a minute, Thurgood, she asked a fair question. What is the state of the union? Where are the rights of the disenfranchised? Did the courts give with one hand and take away with the other? Didn't they force integration in the schools and dismantle American apartheid piece by piece, then take away the tools for full empowerment by endorsing the concept of reverse discrimination by destroying the hope of affirmative action for the poor and providing it for corporate interests? If we were to tally gains and losses, would we come out on the winning side?

Marshall I don't know the answer to that question, but I do know that African Diasporan people are adrift at the dawn of a new century, unable to decide whether success is really measured by buying power. Color is no longer a sign of allegiance and "family" as African American politicians inhabit diversity in ways that we never imagined. Freedom, it turns out, frees one not only from that which oppresses but also from the yokes of race-based allegiances and responsibilities. The worst, of course, is the internecine violence. We are continuing where the Klan and paddyrollers ended. They no longer ride in the night wrecking havoc, death, and destruction; we do.

Yes, there is joy in the community, but without safety, joy is just a temporary and nervous glitch in everyday reality. Joy without safety is more like fear in a clown suit. It may be leaping around in the center ring with ecstatic energy, but when the show is over and the makeup comes off, what have you got? When you want change to occur, it doesn't just rise like bread in the oven without some yeast. During the movement, the violence against unarmed people was the yeast that inevitably caused the people to rise. When they did, King was in place to remind them that "they battled not against flesh and blood, but against powers and principalities." The battle was against the spirit of oppression.

Jordan But laws don't put limits on spirits.

Marshall You're right, but laws can make promises and duties explicit. They offer something to hold on to, something to point the way toward safety, order, and the hope of a better future. Now the law is sometimes like a lever; it's something you can grab and hold while you assure yourself that you are not crazy. You can say to yourself, "Self, this lever is marked liberation and I'm going to hold on to it for dear life until something better comes along." In the midst of lynchings and riots, and every type of civil disorder and neglect imaginable, the law gives you something to grasp.

During the entire time that I was battling against the death penalty, I kept pointing to the rock-solid promise of equal protection under the law. It seldom worked because I was always in the minority, but numbers don't matter when the truth is at stake.

Writer But what is truth? [They looked at each other and laughed.]

Marshall This woman is actually waiting for an answer as if there is one. You go first, Barbara. You tell your truth; then I'll tell mine. [He relaxed, waiting for a good story.]

Jordan You're taking a big chance by letting me go first. When I get done, there might not be any truth left to tell. Let's see, where do I begin? I'm sorry, Thurgood, but the law is more than a lever; it is a marker, a keeper of promises. It is the living heart of the nation and its people. The steady beat of the human heart sets the pace and regulates all of the aspects of the body. The laws that evolve from steady adjudications, precedents, opinions, and public debate infuse the polis with life.

> Law is society's ongoing quest for integrity in interpersonal affairs. It is a sister to the arts and it is something quite beyond ritual and mere liturgy.[7]

Law is creative and constructive. This is particularly true during "interesting times." I lived during extremely interesting times, as the nation was losing its naïveté, when the law was speaking to the nation and the world as an active arm of liberative justice. It was also a time when this country began to embody the elements of the diverse national community that the founders envisioned.

During my lifetime, I saw great progress and great malaise. Looking back on the Civil Rights movement and Watergate, it seems as if it was nonstop excitement and activity, and that is just not the case. Rather, there were long periods of time when little or nothing happened, times when we seemed frozen in a tableau of desire and confusion. What do you do when the leaders of the liberation movements keep dying too young as bullets find their mark? How do you go forward when the voices in the public sphere are so divergent? Should the next move be black power or the poor people's march or both? What will save the people and heal the land? [Turning to me,] You, my dear, are also living in interesting times.

If I were to attempt to speak about the things that have happened in the ten-plus years since my transition from life to life, it would take all evening. So, I will just talk about those things close to my heart.

Writer Can I make a suggestion? I've been thinking about categories that will frame this discussion. The subjects are so important that I would hate to meander and then miss key issues. I'm going to suggest the following format.

1. *Who Will Sound the Alarm? The Courts and Liberation.* Is the entire liberation project at risk? If so, why, and who will warn the populace and provide guidance as to the path forward?

2. *Tell Me A Story, But Do It Quickly Because the Terrorists and Immigrants Are Coming.* How has the story of liberty changed in an era focused on law, order, and security, and how shall we expand the idea of black/white liberation to include all others?

3. *Liberation and the Cosmos—What Has Law Got to Do with This?* How does law transcend itself, and if it does, can it advance the liberation project?

If these subjects meet with your approval, can we begin? Representative Jordan, who will sound the alarm?

Jordan Well . . . [even her hesitations were laced with great rhetorical flourishes]. If you sound an alarm in the middle of a million sirens, who will hear? Sirens of warning have been coming from minorities, women, impoverished nations, allies, and enemies for decades. They were going off before I went into public life and were still sounding when I made my transition.

Marshall You're right, Barbara. That's why I finely retired and should have done it sooner. They kept telling me that I had to hold the seat on the Supreme Court for symbolic and actual reasons. To be blunt, it was so that the opposition party would not have the opportunity to make an appointment, but what good is it to put one finger in a dike of these proportions? All of society must stand up and emphatically say, "We will have liberty, no excuses, no spin." Otherwise, the courts become an instrument of whim, expedience, and the politics of special interests.

Jordan The first warning that I can remember that changed the public perception about law and liberty came from Nikita Khrushchev when he warned that bombs would not be necessary to dismantle democracy in the United States. The statement was perceived as a threat, but it was a prediction that internal forces would erode our strengths. Khrushchev himself remarked, "I once said, 'We will bury you,' and I got into trouble with it. Of course we will not bury you with a shovel. Your own working class will bury you."[8]

Marshall But that's just the point, Barbara. Warnings from outside are never heeded, and warnings from inside are not heard, so what can we offer from this side of eternity?

Writer If I may interject for a moment, you are our legacy; you are the memories temporarily blocked by the immediacy of daily problems. If you speak, we will hear you.

Jordan The warnings are being sounded in your time as loudly as they were in mine. The trouble is that in every generation liberators become so easily distracted by the struggles for personal survival and power. Because individuals have such a daunting task as they fulfill the calling within, we depend on our collective desires for good to advance the cause of liberation. Despite its failures, the United States is a liberty-loving nation.

Marshall Really! Well, to believe that the Unites States is unaware of its global aggressiveness and its lack of concern for the health and welfare of its own citizens is ridiculous. I use the word *ridiculous*, but I find nothing ironic or funny about "the disconnect" between action and intent. If you send me to fight a war but you don't provide for my care when I'm injured, there is a logical conclusion that can be reached, and a warning to those who are choosing to serve. Sarah Baxter, in an article entitled "Squalor of the Vet's Hospital Shocks U.S.," reports the unsanitary conditions at Walter Reed Hospital in Washington, D.C. One parent said this about the conditions at the hospital: "You couldn't get people to mop the blood and urine from the floor while my son was there with his legs wide open."[9] To add to the indignities, claims are being denied and services are limited. The lack of care for those who sacrifice so much is a warning to others.

Jordan Yes, there certainly are enough warning signs, and I do admit that I am perplexed by the political and social choices that the nation has made during the last decade. My love of country is not a mindless acceptance of harmful policies; rather, it includes an informed and scathing critique of current shortcomings. Here are just a few: the spying on citizens without court approval, the end of habeas corpus as we knew it, and the absurd airport rituals that will not stop a smart and focused bomber. But let's focus on the loudest warning signals of your time: the torture bill and habeas corpus.

Marshall Molly Ivins published an amazing piece on September 27, 2006, entitled "Habeas Corpus R.I.P. (1215–2006)." Let's see if I can get her exact words.

> With a smug stroke of his pen, President Bush is set to wipe out a safeguard against illegal imprisonment that has endured as a cornerstone of legal justice since the Magna Carta. This bill is not a national security issue—this is about torturing helpless human beings without any proof they are our enemies. Perhaps this

could be considered if we knew the administration would use the power with enormous care and thoughtfulness. But of the over 700 prisoners sent to Gitmo, only 10 have ever been formally charged with anything. Among other things, this bill is a CYA for torture of the innocent that has already taken place.[10]

From what I can determine, the bill exposes American citizens to the abuses of government that we have managed to avoid during the centuries since the founders of the Constitution placed the checks and balances in place. It makes me so mad that I can't see straight. Rather than repeat the unrepeatable, let me just let you hear the rest of Molly's article. It is worth quoting in full.

The bill simply removes a suspect's right to challenge his detention in court. This is a rule of law that goes back to the Magna Carta in 1215. That pretty much leaves the barn door open. As Vladimir Bukovsky, the Soviet dissident, wrote, an intelligence service free to torture soon "degenerates into a playground for sadists." But not unbridled sadism—you will be relieved that the compromise took out the words permitting interrogation involving "severe pain" and substituted "serious pain," which is defined as "bodily injury that involves extreme physical pain." In July 2003, George Bush said in a speech: "The United States is committed to worldwide elimination of torture, and we are leading this fight by example. Freedom from torture is an inalienable human right. Yet torture continues to be practiced around the world by rogue regimes, whose cruel methods match their determination to crush the human spirit." Fellow citizens, this bill throws out legal and moral restraints as the president deems it necessary—these are fundamental principles of basic decency, as well as law. I'd like those supporting this evil bill to spare me one affliction: Do not, please, pretend to be shocked by the consequences of this legislation. And do not pretend to be shocked when the world begins comparing us to the Nazis.[11]

Jordan This quote takes me back to our original discussion about the responsibility that all of us have to protect our freedoms. On this issue, Molly takes no prisoners. The silence from reasonable people everywhere is shocking. I can understand that the voices of the underclass and the working poor who are being kept within the economic vise of day-to-day struggle are muted by their perpetual labors and their fatigue, but not everyone is in that category. What of the reasonably well off? Do they not understand the dangers of recent legislation?

Marshall The results of the 2006 election prove that they care, but they are operating as if underwater. There is something that happens after an event like 9/11 that slows the public consciousness to a crawl. I think that the constant warning that the terrorists are coming dulls the ability to respond to even the most heinous actions. I like your titles for this discussion, but the second one, "the terrorists are coming," makes no sense. Most countries grow their own terrorists, and the United States is no exception.

Jordan I don't think that 9/11 can really account for the pervasive social malaise. There were catastrophes just as shocking and dramatic during our era, but the public reacted like a lit brush fire despite the unbelievable sequence of events that seemed to tumble one right after another. Imagine, if you will, an era when national guard members shoot students on a college campus, assassinations become commonplace, of civil rights leaders and presidents and finally the president's brother. Those are not normal occurrences.

Marshall No, they are not, but ours was an era of discontent and opposition to authority and 9/11 happened during an era of studied complacency. In 2001, Americans believed the lie about our national invincibility, and the decisive penetration of that mythological shield was enough to stun the nation into a catatonic state. In its aftermath, I believe that if the current government sent out a call to imprison first-born sons of every citizen in the country, there would still be no outcry.

Jordan So what will it take to shake folks from this semiconscious state? While they are sleeping, the tenets of democracy and freedom are being eroded in ways that may never be recovered.

Marshall It will take another generation perhaps, one without the shock of this event burned into their souls. It may take a return to the vigorous debate and public contestation that fuels political vigilance. It may even require an approach to court appointments that views judges as public servants rather than as political functionaries.

Jordan You are not asking for much, are you?

Marshall Now you know that I want the whole thing: peace, justice, and a vibrant public arena. I have never stopped believing that this was possible.

Jordan I agree that this has always been possible, but what is needed is the clarion call, the lone voice crying in the wilderness, the unlikely leader who refuses to give up her seat on the bus. We know what must be done; even when we don't exactly know how to do it, we must try.

Marshall We are getting off course. People are afraid and Roosevelt was wrong. We have a lot more to fear than fear itself. We have to fear those who manipulate and exacerbate our fears. It's one thing to be afraid because of a catastrophic event; it's another thing to be given a face to hate, a focus for our fears.

Jordan The narrative of the nation is one of struggle and overarching principles. We often get things very wrong, but the pendulum of justice swings and eventually prevails.

Marshall Why am I reminded of Edgar Allan Poe's "The Pit and the Pendulum"? The pendulum of justice that you describe is no longer an innocent rhythmic timekeeper; it is a swinging guillotine.

Jordan What a garish thought! What it all boils down to is whether or not there is a single definition of liberation. The question is whether the courts can enforce the liberty that the laws purport to protect.

Marshall Laws can only create the opportunity for peace and justice, and this is no easy task when fear is running rampant. Fear is a greater enemy because it poisons the well of reason with the acids of self-concern. I'm still stunned with the lack of readiness to address the 9/11 attacks. How could these planes stay in the air for so long without being intercepted or shot down? And don't look at me like that, Barbara [speaking directly to me]. You know that there are no conspiracy theories up here. When you can see everything, there is no need for guessing. But just because we know doesn't mean that we can sort things out for you. To do so would affect your destiny.

Jordan The question isn't really how it all ends or where blame lies; rather, it is who should we fear, if anyone. And unless I miss my guess, we're told to "fear" God. Now there is an enemy that you don't want to have. Human enemies are never permanent opponents; they are situational and perspectival.

Marshall But when an attack occurs on your own land, then nothing is reasonable anymore.

Jordan So what should be the response of our nation when planes crash into buildings? Is our republic a viable, God-fearing nation only when things are going well? I don't think that we can talk about a shining nation on a hill and retaliate with such fierceness that we destroy nations that were not even involved in the attacks.

Marshall But it isn't just "the terrorists" that have the nation hostage to its own fears; it is also the influx of immigrants. Barbara, you had a lot to say about that issue, because your state was one of the major crossing points.

Jordan Yes, I set forth a pretty hard line. I wanted identity cards and strengthening of borders, the whole nine yards, but my thinking has changed. We are affecting lives all over the globe. It isn't just a matter of enforcing borders when our economic policies are affecting the lives of those crossing, NAFTA didn't help. It gave American corporations the right to move six inches across the border to make cheap goods, to avoid labor and environmental laws, and to hasten the flow of people trying to escape poverty.

Marshall The way that I see it, the influx of immigrants from Mexico is almost a reverse underground railroad. Those crossing our borders want a better life. There are many stories as to why people cross. Americans believe that Mexicans and Central Americans want to stay in the States, when many of them just want to work and return home. Since the subject is liberation, perhaps the nation needs to consider how our phobias about difference shape foreign policy

Jordan We have to be honest about the concerns. Discomfort with immigration increases when difference is a key factor. We can seal off borders and restrict immigration and we will not be able to preserve a republic that is predominantly Anglo. We are a nation of immigrants and will soon have a majority of Hispanic citizens. This fact should be the source of great delight. The more diverse the body politic, the more opportunities for new ideas and new leadership. If I were to make any suggestion to the future leaders of this nation, I would say look to the neglected segments of society for new ideas.

Marshall You're right, Barbara. If this experiment in diversity is to succeed, it is going to take all of us, without respect to class, economic resources, gender, or sexual preference. We look to the law to frame liberation because both liberation and law have an element of mystery to them.

When well-meaning people assemble to devise systems of order and then those systems take on a life of their own, there will always be an element that transcends the original idea.

Jordan I think that this happens because freedom is God's idea.

Marshall Yes, but liberation also requires great sacrifices of those of us who most seriously seek it. How about this, Barbara? Suppose, just for argument's sake, that we consider the law to be a reflection of the order of the cosmos. Although there is chaos and synchronicity, there is also the potential for creative genesis.

Jordan I remember reading the work of Teilhard de Chardin, a Jesuit priest, mystic, and paleontologist who did a good deal of work on consciousness and the laws of the universe. I think that the connection between law and cosmology can be found in a philosophy of emergence and the recognition that the connections between law and liberation are not necessarily direct or objective.

The laws of nations give clues as to the state of mind of a populace, and sometimes they provide a history of our processive movement toward our highest good. That's all of the science that I know. But from what I understand, there are laws of the universe as well as laws of nation-states. Matter and spirit are intertwined so that "the quanta of matter and spirit that once permeated the early universe become fibers of matter influenced by gravity and threads of spirit drawn by love."[12]

Marshall Where did that come from, Barbara? You didn't learn that in law school.

Let me say a few cosmological things. While our laws are in place to prevent, proscribe, and punish, the laws of the universe seems to be focused on connection, attraction, and a cosmic holding mechanism.

Jordan Yes, listen to this. This is the work of the amazing Teilhard. He wasn't enlightened about all issues, but he certainly offers a unique approach to the nature of reality.

> Crimson gleams of Matter, gliding imperceptibly into the gold of Spirit, ultimately to become transformed into the incandescence of a Universe that is Person—and through all this there blows, animating it and spreading over it a fragrant balm, a zephyr of Union.[13]

Writer Wow, so Teilhard identifies an order that is not static but processively moves us toward a future good.

Marshall Where was Teilhard when I needed him? The idea that we are connected to a future good, and moving toward something better, would have been a breath of fresh air just before I made my transition. You do get discouraged when you struggle for justice all of your life and then observe a return to the status quo. This is not personal, but when I was being replaced by Justice Clarence Thomas, I did wonder about whether my story and the larger story of the nation and the universe made any sense. I wondered if my vision of liberation had been so influenced by the struggle to end racism that I missed a much more powerful form of freedom.

Jordan It wasn't you, Thurgood. All of us, including Constance Baker Motley, understood the political sabotage that inspired the Thomas appointment to the Supreme Court.

> For those of us who are established black members of the legal community, Bush's appointment of Thomas was a stunning rebuff to our painstaking, carefully constructed constitutional theory and a staggering blow to our rise in the profession since Brown. . . . Bush's appointment of Thomas to that Court as the successor to Thurgood seems . . . the most cynical move made in the area of race relations since Plessy. . . . Bush was trying to win an election . . . but at what price to American society and its experience with racism? Bush apparently believed he was retaliating against blacks for deserting the party of Abraham Lincoln, the Great Emancipator, and that such is fair in political wars. In retrospect, there is both grief and boomerang: the black community has been deeply wounded by Bush's misguided calculation, but Thomas has been equally wounded. When the established legal community testified against Thomas at his Senate Hearing, he was grievously wounded personally. No one wants to be publicly cast out by his own people on prime-time television—such a wound even time cannot heal.[14]

Marshall I grieve for him and the weight that he unnecessarily carries. The desire to establish his own uniqueness in contrast to the black community is a fool's project. He will always be a part of the black community. The more that we reject him, and the more that he rejects us, the more apparent are the deep unbreakable bonds of culture, memory, and community legacy. No one escapes the wholeness of this life-giving cosmos, which knits us into the fabric of its past, present, and future. Now that I am on this side of the continuum, I'm certain that the trajectory of human life is toward mutuality and care of self and neighbor.

Jordan Laws that have those principles undergirding them will reflect the highest ideal of the nation's founders. Would you be willing to use the word *love* as a means of describing liberation's connective tissue? Teilhard did. He wrote this lovely small book that addresses the laws of love. He describes love as a holding mechanism; gravity becomes the divine embrace that keeps us from flying off of the planet, and our skin becomes a blessed wrapping that does not contain who we are but keeps our innards in place. We are urged not to pass more laws but to become aware of laws that the universe has in place that protect and challenge us.

Marshall Now that's the expansive view of law that we want to leave for future generations. Sure, I would agree that laws should reflect a philosophy of love, and Teilhard's cosmological view of that word certainly opens up possibilities that I never considered before. I think we have something here, Barbara. "What's Love Got to do with Law?"

Jordan Everything and a bit more!

Summary: The Questions of Law, Liberation, and the Issue of Trust

Writer What a fascinating conversation, yet I'm left with questions as to what Jordan and Marshall would have us do. They were insiders who jousted with the system but ultimately worked within its boundaries. No matter how hard one tries to remember the folks in the "hood," power inevitably erodes the visceral aspects of poverty and its consequences. To arrive at any semblance of liberation, racial/ethnic people who live in the United States must negotiate the complexities and contradictions of a purportedly color-blind society. It is a society that presumes that the law is a protective social tool that benefits all citizens who abide by its precepts, when the facts indicate otherwise.

Sarah Speaking from experience, I know that African Diasporan people have another experience of the law that has little to do with love or liberation. Let me share a few exemplary generalizations. Wise mothers from racial/ethnic communities teach their children to use extreme caution if stopped by the police. They teach their children that it is risky to rely on the "presumption of innocence" and that keeping out of the reach of the law is the wisest and safest thing to do. Anglo families teach their children to trust the police and "the system" because those pillars of society are in place to safeguard their power. What does liberation mean in such a context?

Writer Well, those certainly are generalizations that probably wouldn't hold up to statistical review, but I know what you mean. The social and legal context of this survival guide for the next generation gives us a clue. The fact that segregation was encoded into the laws of the United States and enforced by the courts ruptured trust in ways that have been passed down from generation to generation. And yet, it was the law that ended segregation and integrated schools.

Sarah One of the overarching questions that emerges during this conversation between Jordan and Marshall is whether a marginalized community moving toward a participatory and reconciling relationship with a dominant culture can or should reorient their relationship toward the judicial system.

Writer Since we are beginning to realize that a color-blind society may not be a reasonable or desirable goal, how then shall we negotiate difference and reframe "group image"? What the society thinks of a group as a whole (deeply sedimented stereotypes) contributes to the rulings of juries and the split-second decisions of police officers. There is a certain comfort in summarizing our knowledge into sound bites—most Asians are smart, most blacks are criminal, most whites are law abiding, and so forth.[15] No one does this consciously, but such faulty information can be critical when a police officer must decide whether to shoot or pause when he sees a dark object in a racial/ethnic person's hands. Is it a wallet or a gun? How an entire ethnic group is perceived can determine the life or death of an individual who bears those immutable traits.

Sarah These conversations are not definitive. They mark the beginning of dialogue. Jordan and Marshall seem to be urging the next generation to continue the struggle on all fronts. They don't want the next generation to cede areas of law and justice just because those areas have not always benefited the community. There is something about liberation that cannot flourish in chaos. Liberation can emerge from the creative genesis of the unexpected, but ultimately it must translate its most ethereal desires into pragmatic action.

Writer I am hoping for a pragmatism infused with an openness to the unknown.

The Exercises

1. Discuss your experiences with the law, positive and negative. Describe the opportunities for liberation in a society based on law. How do your answers affect applications of law to, for example, Mexican border crossings, alleged Islamic terrorists held at Guantanamo Bay, and Appalachian families captive to the corporate interests of the coal mining industry?

2. Name African American role models in public life. Is there a relationship between their success and the liberation of marginalized members of their communities?

3. What role does the media play in the formation of "deeply sedimented stereotypes"?

4. Describe the ways in which the law is a help and a hindrance to minority communities in the twenty-first century.

For Further Reflection

Thurgood Marshall

Aldred, Lisa. *Thurgood Marshall: Supreme Court Justice*. Philadelphia: Chelsea House, 2005.

Baine, Kevin T. "Wit, Wisdom, and Compassion of Justice Thurgood Marshall." *Hastings Constitutional Law Quarterly* 20 (Spring 1993): 497–502.

Ball, Howard. *A Defiant Life: Thurgood Marshall and the Persistence of Racism in America*. New York: Crown, 1998.

Barker, Lucius J. "Thurgood Marshall, the Law, and the System: Tenets of an Enduring Legacy." *Stanford Law Review* 44 (Summer 1992): 1237–47.

Bland, Randall Walton. *Justice Thurgood Marshall, Crusader for Liberalism: His Judicial Biography*. Bethesda, Md.: Academica, 2001.

———. *Private Pressure on Public Law: The Legal Career of Justice Thurgood Marshall*. Port Washington, N.Y.: Kennikat, 1973.

Bloch, Susan Low. "Thurgood Marshall: Courageous Advocate, Compassionate Judge." *Georgetown Law Review* 80 (1993): 2003–9.

Davis, Michael D., and Hunter R. Clark. *Thurgood Marshall: Warrior at the Bar, Rebel on the Bench*. Secaucus, N.J.: Carol, 1994.

Fenderson, Lewis H. *Thurgood Marshall: Fighter for Justice*. New York: McGraw-Hill, 1969.

Goldman, Roger L., with David Gallen. *Thurgood Marshall: Justice for All*. New York: Carroll & Graf, 1992.

Greenberg, Jack. *Crusaders in the Courts: How a Dedicated Band of Lawyers Fought for the Civil Rights Revolution*. New York: Basic, 1994.

Marshall, Thurgood. *Supreme Justice: Speeches and Writings*. Philadelphia: University of Pennsylvania Press, 2003.

O'Connor, Sandra Day. "Thurgood Marshall: The Influence of a Raconteur." *Stanford Law Review* 44 (Summer 1992): 1217–20.

Rowan, Carl Thomas. *Dream Makers, Dream Breakers: The World of Justice Thurgood Marshall*. Boston: Little Brown, 1993.

Tushnet, Mark V. *Making Civil Rights Law: Thurgood Marshall and the Supreme Court, 1936–1961.* New York: Oxford University Press, 1994.

———. *Making Constitutional Law: Thurgood Marshall and the Supreme Court, 1961–1991.* New York: Oxford University Press, 1997.

———. *Thurgood Marshall: His Speeches, Writings, Arguments, Opinions, and Reminiscences.* Chicago: Lawrence Hill, 2001.

Williams, Juan. *Thurgood Marshall: American Revolutionary.* New York: Times Books, 1998.

Barbara Jordan

Aghahowa, Brenda Eatman. "Grace under Fire: The Rhetoric of Watergate and Patriotism, Barbara Jordan Style (Texas)." Ph.D. diss., University of Illinois at Chicago, 2004.

"Barbara Charline Jordan," in *Black Americans in Congress, 1870–1989.* Prepared under the direction of the Commission on the Bicentenary by the Office of the Historian, U.S. House of Representatives. Washington, D.C.: Government Printing Office, 1991.

"Barbara Charline Jordan," in *Women in Congress, 1917–2006.* Prepared under the direction of the Committee on House Administration by the Office of History and Preservation, U.S. House of Representatives. Washington, D.C.: Government Printing Office, 2006.

Blue, Rose, and Corinne Naden. *Barbara Jordan.* New York: Chelsea House, 1992.

Bryant, Ira Babington. *Barbara Charline Jordan: From the Ghetto to the Capital.* Houston: Armstrong, 1977.

Canas, Kathryn Anne. "Barbara Jordan, Shirley Chisholm, and Lani Guinier: Crafting Identification through the Rhetorical Interbraiding of Value." Ph.D. diss., University of Utah, 2002.

Crawford, Ann Fears. *Barbara Jordan: Breaking the Barriers.* Houston: Halcyon, 2003.

Fenno, Richard F. *Going Home: Black Representatives and Their Constituents.* Chicago: University of Chicago Press, 2003.

Green, Robert L. *Barbara Jordan: Daring Black Leader.* Milwaukee: Franklin, 1974.

Haskins, James. *Barbara Jordan.* New York: Dial, 1977.

Holmes, Barbara Ann. "Barbara Jordan's Speeches, 1974–1995: Ethics, Public Religion and Jurisprudence." Ph.D. diss., Vanderbilt University, 1998.

———. *A Private Woman in Public Spaces: Barbara Jordan's Speeches on Ethics, Public Religion, and Law.* Harrisburg, Pa.: Trinity Press International, 2000.

Horwitz, Linda Diane. "Transforming Appearance into Rhetorical Argument: Rhetorical Criticism of Public Speeches of Barbara Jordan, Lucy Parsons, and Angela Y. Davis." Ph.D. diss., Northwestern University, 1998.

Jacobs, Linda. *Barbara Jordan: Keeping Faith*. St. Paul, Minn.: EMC Corporation, 1978.

Jeffrey, Laura S. *Barbara Jordan: Congresswoman, Lawyer, Educator*. Springfield, N.J.: Enslow, 1997.

Johnson, Linda Carlson. *Barbara Jordan, Congresswoman*. Woodbridge, Conn.: Blackbirch, 1990.

Jordan, Barbara. *Speaking the Truth with Eloquent Thunder*. Edited by Max Sherman. Austin: University of Texas Press, 2007.

Jordan, Barbara, and Shelby Hearon. *Barbara Jordan: A Self-Portrait*. Garden City, N.Y.: Doubleday, 1979.

Kelin, Norman, and Sabra-Anne Kelin. *Barbara Jordan*. Los Angeles: Melrose Square, 1993.

Kirk, Rita G. "Barbara Jordan: The Rise of a Black Woman Politician." Master's thesis, University of Arkansas, 1978.

McNair, Joseph D. *Barbara Jordan: African American Politician*. Chanhassen, Minn.: Child's World, 2001.

Mendelsohn, James. *Barbara Jordan: Getting Things Done*. Brookfield, Conn.: Twenty-First Century Books, 2000.

Rhodes, Lisa Renee. *Barbara Jordan: Voice of Democracy*. New York: Franklin Watts, 1998.

Roberts, Naurice. *Barbara Jordan: The Great Lady from Texas*. Chicago, Ill.: Children's Press, 1984.

Rogers, Mary Beth. *Barbara Jordan: American Hero*. New York: Bantam, 1998.

Thompson, Wayne N. "Barbara Jordan's Keynote Address: The Juxtaposition of Contradictory Values," *Southern Speech Communication Journal* 44 (Spring 1979): 223–2.

The black struggle for freedom is at its heart a profoundly human quest for transformation, a constantly evolving movement toward personal integrity and toward new social structures filled with justice, equity and compassion.
—Vincent Harding

4. A Liberated and Luminous Darkness: Spirituality of Struggle

The little space within the heart is as great as this vast universe. The heavens and the earth are there, and the sun, and the moon, and the stars; fire and lightening [sic] and winds are there; and all that now is and all that is now: for the whole universe is in . . . [the Divine One] and . . . [the Divine One] dwells within our heart.
—Matthew Fox

When your heart breaks,
The universe can pour through.
—Joanna Macy

In this chapter, Rosa Parks and Howard Thurman are the conversation partners. They share their thoughts about the power of contemplative practices to influence the moral flourishing of future generations. They discuss contemplative resistance and restoration as a means of bridging cultural gaps, and they ponder the cosmological version of Thurman's hope for interfaith cooperation.

The Elders

Rosa Parks, 1913-2005

As long as people use tactics to oppress or restrict other people from being free, there is work to be done. Although we have made many gains, racism is still alive.

My message to the world is that we must come together and live as one. There is only one world; and yet we, as a people, have treated the world as if it were divided.[1]

When it comes to liberation, Rosa Parks is a national icon now, but her early life in Tuskegee, Alabama, was humble and endangered. Born in the deep South, she lived within the constraints of segregation and graduated from high school in 1934. In 1943, she became secretary of the Montgomery, Alabama, NAACP. On December 1, 1955, on an ordinary day, she boarded a bus to return home after a long day working as a seamstress in a department store. Before the ride was over, she had refused to give up her seat in the "colored" section of the bus and was under arrest. The story that she was tired is true, but it is also true that she was prepared. Having been trained at the Highlander School, and active in Montgomery with the NAACP Youth Council and voter registration, she was ready.

Sometimes liberation comes with a bang because a wall falls or a dictator dies, and sometimes the act that frees people is as gentle as a whisper or a resolute "no" to segregated buses in Montgomery, Alabama. The story usually ends with the arrest, but Parks stood trial, was found guilty, and lost her job. In 1957, she moved to Detroit. During her lifetime, she received an honorary doctorate degree from Soka University and a Rosa Parks Peace Prize.[2]

Howard Thurman, 1900-1981

Teach us to overcome our fear of life; and in that freedom may we learn to understand life and, in our understanding of life, to love.

Howard Thurman was born in segregated Daytona Beach, Florida, with an inclination toward the mysteries of the cosmos. As far as Thurman was concerned, deep spirituality was not just the result of birthright and folk interpretations; he was also highly cognizant of his place in a broader universe. In *With Head and Heart*, Thurman recalls seeing Halley's Comet and feeling connected to something more profound than the petty social arrangements that relegated him to the margins of society.[3] It was this connection to something more that inspired Thurman's approach to inner liberation as an inextricable aspect of public liberation. Freedom, in Thurman's thinking, is tied to knowledge of self and knowledge of a free God. Liberation comes from a release of fear. Fear can severely curtail human responses to the risks inherent in daily life.

Thurman was ordained in the Baptist church and served in the ministry as pastor, chaplain, professor, and founder of the Church of the Fellowship of All People. Despite these official responsibilities, he is better known as a mystical thinker, a spiritual resource for the Civil Rights movement, an advisor to Martin Luther King Jr., and a proponent of social and spiritual unity.

The Conversation

After the first interview, it has become clear to me that what I bring into the discussion by way of my own thoughts and priorities has a way of turning up in the conversation even if I never say a word. So, given the fact that I am about to interview one of the preeminent mystics of my time, it would probably be prudent to admit that there is something about mysticism that will not let me shift my gaze. I am drawn by its siren call during the most ordinary of circumstances. By mysticism, I mean the interpenetration of multiple realities that testify against deterministic and objectified presumptions about the life space.

A simpler way to say this is to admit that things happen that we don't understand, and that there is wisdom beyond the limits of our personal knowledge. The elders knew this and tapped their feet and nodded their heads when words failed. If what we can see and touch is all that there is, then what do we do with the witness of the ancestors who refused to draw immutable lines between life and death, between the future and the past? What do we do with the foundations of faith that always extend beyond the soil of current lives? The roots of faith emerge from a vibrant and accessible past and spread toward a future that is created with each decision and each breath.

Always we are reaching toward a God who inhabits that totality and beckons us from fragmentation toward wholeness. I hate to be the one to break the news, but liberation is a very mystical idea. This statement makes me feel as if I am yelling "Fire!" in a crowded theatre because everyone around me seems to believe that liberation is a goal that is tangible and fixed. There is the unstated presumption that the underprivileged and oppressed will "overcome" someday and will attain a state of completion that we will then dub liberation.

I have no such hope. For me, liberation has always been as slippery as an eel, an electric eel that will shock you if you don't handle it just right. Freedom changes costumes and transmogrifies in ways that make containment impossible. Just when you think that you have described its manifestations and have a firm grip on the methods and practices that will lock its benefits in for all time, it escapes. Even in the places where liberation battles were fought and won, it is not an established reality. In fact, in South Africa and America, where the malevolent twins apartheid and segregation once prevailed, one finds in place of defiant hierarchical structures of oppression a more elusive form of white privilege and the status quo.

The fact that some things are better does not add up to liberation. For decades, we have been preaching and marching and legislating toward liberation and missing the mark each time. The reason seems to be our failure to understand the spiritual interiority that is the essence of liberation. Liberation is a wisp of self-reflection, the refusal to surrender self-respect for survival. It is a framework that is open enough to construct temporary dwelling places, but ultimately liberation is a goal that hastens ahead of us.

Could Rosa Parks and Howard Thurman help with any of this? I admit that I had some concerns about this interview. In life, both Thurman and Parks seemed to have a formality about them that they wore as a shield and a defense against a dangerous society. Rosa seemed to be a practical, no-nonsense woman who might not appreciate my ethereal approach to liberation. Thurman, on the other hand, was a visionary whose approach to freedom inspired leaders and laity alike. Through his work, he became my spiritual mentor, not so much for his religious eclecticism as for the power of his poetic mysticism. This was going to be good.

I was looking for the white-haired, braided symbol of the Civil Rights movement and was surprised when Rosa entered the room sporting dark hair and a light heart. When Thurman arrived, they greeted each other warmly. Such relaxation was unexpected. The measured elocution was gone, and Thurman's voice lilted with each word; Rosa, so staid in her resistance on the bus and so regal in her senior years, now exhibited the same characteristics, but with an informality that was refreshing. Before I could set the agenda for the conversation, Thurman began.

Thurman Let's get right to the point. We are having this meeting because the generations that followed us have forgotten the original goals. The ends got lost in the means and methods. Am I right about that?

Parks I'm not sure that we had specific goals, other than the desire to unbend our backs and reclaim our identity. Still, you may be right about the fact that folks seem to have forgotten why we struggled. I just can't believe that their memories are so short. I could understand that type of amnesia after a few centuries. But you all are less than fifty years from the events that changed the lives of African Americans in North America forever and launched liberation movements throughout the world.

Thurman But it really doesn't take long, Rosa. The lures of materialism are powerful cultural signifiers. When we are allowed to splinter the concept of divine presence into fragmented characteristics, we end up with idols made in our own image. I have always believed that the acknowledgement that there is one God, one Spirit, dismantles fiefdoms built on the foundations of false difference.

Parks So, what you are saying is that in one or two generations we have lost the incentive to attain certain goals as a community, in favor of individual achievement. If that is the case, why isn't that individualism expressed in a desire for education? There is a rampant anti-intellectualism among young people that is exacerbated by the entertainment industry.

Thurman I love to hear you speak your mind, Rosa, but I'm not certain that I agree. To be anti-intellectual, one has to consider the thing being rejected. Instead, there seems to be a lack of interest entirely. The generation that stands on the cusp of leadership watched the boomer generation effect massive social change but short-lived personal gains. Ultimately, many of the boomers devoted their lives to upward mobility. Certainly, the generation right after the Civil Rights movement went to school and obtained degrees, but at what cost? If you demand to be educated by the people who were oppressing you just moments ago, can you be certain that oppression won't be seeded in their pedagogy?

Parks I don't know the answer to that question, but I suspect that integration is going to get blamed for the current ills.

Thurman Not at all, Rosa. Integration was a crucial demand that highlighted the disparities between the races and the constitutional entitlement of all people to equal treatment under the law. It's important to talk about integration as fair access to God's creation. But if liberation is built on the premise that I must sit next to you to learn, if I must be taught in Eurocentric institutions and adopt Eurocentric values to the exclusion of my own, then I have lost something intangible and irreplaceable.

Parks I can't argue with that, Howard. Whether we prefer integration or the intentional, communally based teaching of our own children, there has to be a spiritual center that energizes those efforts. My grandmother would have said, "There has to be something on the inside." Liberation has to take on the shape of our lives or it won't fit. You know, I hadn't thought of this before, but if liberation is dropped on the community like a mantle, it will be hard to recognize the change. Liberation has to arise from a context of centered transformation. When structural oppression is removed, the people must be given the opportunity to grow into their freedom at their own pace. This race for the finish line doesn't work if the goal keeps moving.

It's hard to describe this because it is coming to me as I speak. But the problem seems to be that we presumed that segregation had no long-lasting effects. We presumed that if we could dismantle legal and social separation, then we would be able to excel. I lived long enough to realize that oppression leaves a psychic wound that must be healed before mobility can begin. One does not walk out of Auschwitz, dust oneself off, and say, "Let's get on with it." If you do, there will be a day of reckoning when the walled-up grief explodes within your soul.

Thurman Yes, Rosa, and grief can manifest in any number of ways. Africans in the North American Diaspora have become the ultimate consumers of material goods. They are drowning in debt and succumb to consumer lusts that seem impossible to satisfy. I believe that this is a way of laying claim to a liberation that was pronounced but never constructed from the inside. It is also a function of lost identity. Liberation is not a one-size-fits-all benefit. If we don't know who we are after years of denigration, then liberation won't set down deep roots of self-understanding in the people.

Parks Is there a cure for this lack of self-understanding? Living beyond your means is a dangerous form of denial. The other evidence of denial is the antagonism of black people toward new immigrants from Mexico and other areas of the world who are being exploited for their labor as we once were.

Thurman I'm not sure that the exploitation of labor is over for the black community, and I'm not certain that black people have any real animosity toward immigrants; they are just angry about being displaced by newcomers and fearful of being permanently on the bottom of America's social hierarchy. That, of course, is no excuse, and I want to advocate for an ethos of mutuality in liberation.

Parks How are we going to talk about mutuality when the entire liberation movement is stalled?

Thurman I agree that the movement that began in the sixties is stalled, but should we be trying to salvage a forty-year-old vision of liberation? I want to move toward a twenty-first-century model. I want to give some guidance as to how to plot a new pathway toward wholeness and healing. Toward that end, my work on the inward sea may have been more prophetic than I realized. One can never focus completely on social accomplishments without careful attention to the well-being of the spirit. It is the strength of the inward that ensures long-standing success of social and political agendas. It is the power of interiority that sustains during oppression and enlightens during moments of freedom. Although rampant consumerism seems to extinguish that interiority, I don't want to end up blaming the victims for their entanglements in capitalism's snares.

Rosa Neither do I, and yet I have to say what's in my heart. Earlier, you spoke of the loss of goals. It isn't that we didn't know what we wanted to accomplish through sit-ins and protests; it is that the assassinations fragmented the spiritual and intellectual capital of the movement. No matter how much inward strength you possess, the shattering of hope and confidence will cause common purposes to implode. Then the only path seems to lead toward personal rather than communal empowerment. We are seeing the results of survivalism. Each death of a communal leader was a message to the masses to keep their heads down and fight for self and self only.

Thurman That explains the private Jesus in private hearts instead of the lay priesthood of believers acting as the body of Christ to the world. The other issue is that one can hardly prepare for the aftermath of resistance while in the midst of active resistance. I'm going to choose a biblical example to make my point. I'm thinking of King David now, running from Saul, fighting with enemies, losing sons, not to speak of the lament of the women as they watch the men playing deadly war games in the name of God. It's all too much, and to survive one must stay in the moment with focus and strength. Foresight and view require historical distance. You can't have a wide-angle view while you are in the fight.

Parks True. While marching against racism in Birmingham, one cannot make a mental note to avoid the pitfalls of consumerism in the future. It doesn't work that way. During the height of movement activities, the tasks before us were too formidable. We faced the behemoth of American injustice and could only keep our hands on our slingshots, our songbooks, and our three smooth stones. We also believed with all of our hearts that there would be time after the bastions of America's apartheid fell to reconstruct priorities and to have public dialogue about communal goals and how we would achieve them. That time never came.

Thurman Yes, all of that is true, Rosa, but an equally important point is that the African Diasporan community no longer remembers its collaborative style of leadership. The authority of elders has been replaced by the model of a single charismatic leader (made in the image of the Christ who resembles a European king more accurately than a Hebrew prophet).

During the Civil Rights movement, the people desired a king, and they got one. Martin Luther King Jr. became a reluctant spokesperson for their collective desires for liberation. When you don't know whether or not you will survive the crushing power of segregation, you must have a point of focus. MLK provided the critical voice and face of resistance. After his death, the lack of centralized leadership created a void that seems to have been filled with the desire to acquire. In one generation, everything that had been denied to all previous generations was obtained on credit. The end result was that involuntary slavery ended and, for some members of the community, economically based voluntary slavery began.

You know, Rosa, voluntary bondage is the most insidious form of reenslavement because one cannot march or sit in to oppose the rapacious needs and desires of the human spirit. One cannot use passive resistance against unacknowledged personal greed.

Parks [turning to me] Now that we have entered the realm of the ancestors, we really do know what is going on beyond the earth life. While we do not spend time sitting around cosmic televisions, there is a knowing that encompasses all that occurs. I say this so that you will understand why I know about a television show about sixteenth birthday parties for the rich and not so rich. I think that it comes on MTV, and it is shocking—the crass materialism, the expenditures, the "bling" at sixteen, the tantrums, the parents held hostage by children and their lack of spiritual focus. Their only concern is about the big name who will show up at the party.

Thurman Rosa, you have chosen an excellent example of the voluntary enslavement to materialism, but I think that you are wrong, when you say that this materialism has no spiritual focus. Consumerism is a form of worship. They are worshiping the gods of technology, trinkets, and toys. I think that the creation of the middle class (a designation that has less and less meaning every day) required not just a specific work ethic and description of values but also explicit and implicit dictates as to how people should live, what they should desire, and also what (rather than who) should be worshiped.

Spending money is a way of proclaiming a freedom that does not exist. The truth is that the economic formula of Western civilization that has been exported all over the globe is a form of slavery, and this type of slavery is insidious and more dangerous than chains. In this system, everyone works for the market economy. The compensation is never commensurate with the required labor; instead, workers are offered a mirage of credit-based purchasing power that keeps everyone compliant.

Parks At the Highlander School, and through Martin's bravery, we learned how to resist violence. Our experiences in slavery and the wisdom of the elders taught us to survive, but it is just as important to guard our spirits from our own desires. I really believe that the churches are complicit in the desire for consumer goods.

A silence descended as Parks and Thurman pondered the state of things in the twenty-first century. I waited for someone to say something and then began to fidget.

Writer Could I suggest some topics for discussion so that we can leave with your wisdom for future generations?

Thurman The wisdom for future generations is in my books, but I noticed something just now. You just interrupted a rich moment of contemplation with a question. Meditative moments are not empty; instead, they provide some of the richest and most healing opportunities for respite and spiritual insight that we can inhabit. Although contemplation need not be silent, silence is the ultimate luxury of the human spirit. But the same effect can be achieved through the cessation of normal activity with or without silence. Rosa, the power of your resistance on the bus was enhanced by your contemplative resolve. To make a decision and then to act not with debate or argument but with purpose and integrity is powerful beyond words.

Parks It looks as if we are heading in the same direction. One should stop to consider the lay of the land, to consider the potential for pitfalls and to determine where you are on the journey. To hurl headlong toward freedom without an understanding of history and without the cultural wisdom passed down through the generations may be more dangerous than slavery.

What does it matter if you can go to integrated schools and participate fully in society if you hurl yourself off of the precipice of family dissolution and persistent self-loathing? What does it matter if the generation that wandered in the wilderness can finally buy a flat-screen TV on installment payments that require so much overtime that the youth die of neglect, drugs, and gang warfare?

Thurman Power has many manifestations, but for emergent communities it should be modeled and used to advance moral precepts, community building, and self-knowledge. All of the external education in the world will not help us to know ourselves. Emotional eating, sports and television addictions, workaholic behaviors, frenetic worship services, and prosperity preaching are just a few indicators that there is a need for internal peace.

There is more to be said, but just for a moment let's just listen to our breath and imagine the gift of vitality that is offered each time we inhale and the opportunity for renewal that is offered each time we exhale the toxins that build up in each cell. Our bodies, whether healthy or not, are finely tuned instrument with capacity for more than shopping and entertainment. How then shall we thank the God of the universe for bringing us through, for bringing us out and refreshing us with every breath?

The silence lasted longer this time. I was startled by the depth and rest that it provided. This time, I waited until Thurman signaled the end of the meditation by getting up.

Writer OK, shouldn't we conclude with an analysis of power and inner strength? Both are needed to restart the liberation initiatives.

Parks First, don't we have to redefine liberation? When I refused to get up from the bus seat, when Martin marched and Malcolm railed against the artificial constraints of segregation, it was not to grant a small sliver of freedom to earthbound people. It was the spiritual launch of a liberation too vast to be circumscribed by a single life. This is a liberation worth dying for, worth risking everything for.

The narrative I hear too often is that there is a moment in time when African Diasporan captives were liberated. Instead, there was a moment in time when the cage doors were opened so that a people who knew how to soar could lift off. The intent was to awaken that which could never be confined in the human spirit.

Thurman The power that is meaningful for future generation comes through the human spirit but emanates from a divine source.

Parks But, Howard, does that mean that although we are both Christians, only those who know Jesus can mediate the power?

Thurman I didn't say that. I said initially that God is a spirit and the oneness and holism of that spirit can be received through whatever cultural, denominational, or faith paradigm that beckons. I admit that this idea was not well received in black churches in my day. The Church of the Fellowship of All People stands as a lonely lighthouse beaming the essence of a great idea of unification of human spiritual endeavor. It is an idea that never caught on. I believe that the wave of the future is beyond the sectarian needs of individuals and toward a manifesto of common spiritual purposes.

If we focus on the particularities, we lose the big picture. For example, the particularities of Africana, Mexican, GLBT, women's rights, HIV/AIDS, and working-class poor of all ethnic origins divide us into camps focused on the most obvious differences and conflicts when, in fact, each liberation movement bears a passion seeded in the principles of fairness, justice, love, and neighborly cooperation. Each of these terms must be rescued from the status of banal platitudes to once again do the work of uniting us.

Parks So, you believe that there are common principles, first causes as it were, that serve as the foundation of liberation movements. When those principles are lost or forgotten or are so deeply sedimented into the public consciousness that we are no longer consciously aware of their effects, it is inevitable that we will lose our way.

Thurman Yes, Rosa, that is what I believe.

Parks I had purposed in my heart not to ask this question, but I have no choice because I really want to know. Did you really believe that we could worship beyond boundaries of cultural, denominational, and faith orientations, and if so, what do you do after 9/11? What does worship look like in the Church of the Fellowship of All People after 9/11?

Thurman It looks exactly as it would after Rwanda, Bosnia, Hiroshima, and Nagasaki, after religious conflicts around the world. The belief that there is a privileged approach to God is a legacy of great religious hubris and a crusader mentality. We all stand before a God who is beyond our explicit knowing. The embodied emissaries and holy progeny of this God attest to this fact. One would expect thoughtful people to be humbled in the face of such vastness, such cosmological intention, but instead we present ourselves before God soaked in the blood of our sisters and brothers.

As long as there are human beings, there will be holocausts. I only suggested that, even in the face of our unrelenting inhumanity, it is better to worship together than to worship apart. Now I am not suggesting that coming together increases our righteousness or our insight into the mind of God, but I am suggesting that it may be the only comfort that we have during our times of greatest pain.

A long period of time passed, before anyone spoke.

Thurman Rosa, tell me what advice you have for the next generation? I've been doing most of the talking.

Parks Well, I am in absolute agreement with regard to common purposes. Wherever people are being persecuted, we have to stand with them.

We can't use a reading of our sacred texts to affirm one people and then castigate another group with the same texts. The claim that we are God oriented does not give us the right to tell others how they should exercise their innate freedom. But let me be clear. Although community is the matrix that we are all born into, there comes to every person a moment to live out beliefs in a concrete way.

At that moment, it is the prophetic task of the individual to stand for what is right in ways that may even jeopardize one's survival. You usually get no warning as to when that moment will come. I didn't. It was always so strange to me that this one moment in time changed so much. I was not compelled to sit; I was offered the opportunity to respond to a life-changing moment. Ida B. Wells-Barnett had the exact same moment that I had, but it was in Memphis. She refused to leave a trolley car and was dragged off by the police. She sued and won on the lower level but then lost on appeal. The time was not right. The hardest thing to do is to wait for the fullness of time. We can't see the complete arc of justice as it bends throughout God's cosmos; all we can do is hear the Spirit and respond when our time comes. The rest is up to God. Because I trust the cosmological view of liberation, I don't think that the slave revolts of John Brown and Nat Turner were failures. Each moment of resistance tilled the soil for the future.

I'm not unusual in this regard. Others have moments and pass them up as ordinary choices. I don't believe that there are ordinary choices. We live in between God's breath and God's grace, and each choice moves us closer or farther from our source of being.

Thurman Now you have me thinking as to whether or not I ever had such a moment.

Parks Of course you did. You described a sense of mystical purpose in Daytona Beach that indicated a perception of something greater that was luring you toward your ultimate destiny. I understand that you were born with the veil and had your ears pierced to let some of that power dissipate. The consumerism that pervades the sensibilities of oppressed people flattens the perception of divine call. From what I can tell, a whole generation seeks external proof of its own worth and at the same time tries desperately not to call attention to itself.

Liberation requires individuals willing to stand when no one else will, to sit when others are threatening you with harm, to embrace an outsider in full view of an insider, to proclaim the wisdom of the ages and the already/not yet justice of God in the midst of horrific circumstances. We do this although we don't know what the end will be, and we do this because liberation is the responsibility of each and every person. I know that the sacred heart of the liberation story lies in ordinary acts of obedience and resistance by ordinary people.

Thurman Thank you for that, Rosa. Liberation is not a goal or an event to be enjoyed. It is a series of events that draw us closer to true liberation in God. Liberation comes in the moment that we hear the leading of the Divine and follow. It is the freedom to unbind the shackled and to reunite with God and neighbor. Until we achieve that reunion, we move from liberation to liberation gathering seekers as we go, celebrating only long enough to encourage our spirits and then moving on to new struggles around old issues in different contexts

Parks I hate for this to be such a love fest, Howard. There has to be something we disagree about, but I must say that your interfaith approach to the "gathering of seekers" adds meaning to the endeavor. You never know who your partners on the journey toward freedom will be. Each partner adds something that undergirds and enhances the situation. The African American struggle for full justice is enhanced when they welcome the issues of Mexican immigrants, because the issue of labor exploitation of the underclass includes not just race but also ethnicity and class.

Thurman We don't get a road map of God's intentions. All we get is the opportunity to live every moment with intentionality and power. The way forward for this generation is toward an awakening of the spirit within, a focus on the good of the community, and prophetic activism of the individual. Prophetic activism is an act of radical obedience because it serves a cause greater than political or social ambition. The prophetic impetus allows commonalities to arise among divergent seekers of justice.

Parks But don't you think that biblical language tends to obscure the ordinariness of prophetic activism? The language of the 1960s was not the language of the 1920s. Each generation encounters God and proclaims in their own voice their astonishment and gratefulness that we are held within the embrace of a cosmic deity. We are not required to put on Grandma's apron or Malcolm's hat to approach God. We get so caught up in predictability and the comfort of sameness.

I was always seen as so proper in my dress and demeanor. The gloves and hat were a uniform of respectability that protected us against accusations of diminished humanity. Now the hip-hop generation and beyond turns that false propriety on its head and dresses in a way that proclaims their liberation from expectations.

Thurman OK, Rosa, I'm not with the draggy pants and torn clothes, but I see your point. If we are free indeed, then churches wouldn't all look alike.

I am praying for a fracturing of the acceptable gloss in worship. How can liberation have teeth in society if the churches teach conformity to a set of acceptable worship behaviors that have nothing to do with God? John the Baptist would not fit in any current congregation that I know of. Yet he heard from God and ministered out of a deep sense of calling that could not be conformed to the religious structures of the day.

Parks But should worship be conformed to popular culture? There seems to be a tendency to follow one formula or another, and there seems to be only a small window for creativity and the fresh wind of the Spirit. How do we hear the Word of God collectively without constructing idols? You know the whispering game where one person whispers to another, and they pass it down. When it reaches the last person, there is no semblance to the original statement. It is as if we have been given wonderful spiritual legacies in various religious faiths, but by the time we pass it down, it no longer resembles anything other than our own intense desire for order and predictability.

Thurman You are right, Rosa! The reason for the distortion seems to be the inability to listen. The contemplative moment creates anxiety in the busy and in the confused. And, all of us fall into one of those categories at some point in our lives. We are urged to pray, and even that command is translated into talk and activity. Just suppose that the mandate to pray without ceasing is the call to inhabit a space of contemplative living that situates a human life in the pause between the leading of the Holy Spirit and the free-will "yes." In this place, there is peace even at the height of the chosen activity.

Parks So, this is not necessarily silence. It is timelessness. The marches during the Civil Rights movement took place in that displacement of time. I remember those days almost as if they were dreams. We were busy, we were moving, but we were "in but not of" the world. Following spiritual guidance can feel as if you are in a parallel universe. We were looking beyond the practical realities of danger and death so that each moment was saturated with a palpable peace. We were not operating out of our own strength; to do so would have been impossible. We had so little strength left.

When the story is told that I was tired, they forget that my weariness was not from eight hours of work but from the exhaustion of generations gone on and the weight of constant exploitation. Most of the victories of the Civil Rights movement were followed or preceded by acts of violence. We could not bask in the success of a march because we were startled back to reality by Emmett Till's murder in August 1955 or the Birmingham

16th Street Baptist Church bombing on September 15, 1963. After each incident, it took everything in us and all of the faith that we could muster to go on. The impetus toward liberation is never easy. Once you start thinking about what you're doing and how difficult it all is, and the fact that you don't know what the outcome will be, the human spirit stalls.

Thurman It's even more difficult when there is a vacuum in leadership. After catastrophic losses of leadership, liberation movements tend to inch along toward their destiny. They don't want to move too quickly, and they don't want to look toward the future, because they are hoping that a leader will catch up with them and yell "Charge!" And when none comes, those halfway down the path must determine how they will continue.

Parks Isn't that where this new generation is now? Waiting and listening, shopping and marching in place? The power of the movement today is hidden within. It is the freedom of saying no to that which is obvious and at hand. Sometimes I'm amused with the characterization of what happened when I refused to give up my seat on the bus. I wasn't in the white-only section; I was in my place in the "colored" section when trouble came to me. When trouble comes to you, there is no option; it must be resisted. The first step toward liberation is the opportunity to refuse to participate in your own oppression. Often liberation is right in your face; you don't have to go anywhere to attain that state of mind, but you do have to say no to the familiar oppressions that have become your constant companions.

Thurman [shaking his head, with a wry smile on his face] Are you telling this new generation to just say no? You have lost me. I never understood the antidrug campaign slogan. How do you tell teenagers in the ghetto to just say no to drugs that are being flown to their doorsteps without their yes?

Parks I'm suggesting a "no" that is not disconnected. Resistance is embodied commitment. There has to be a nurturing of the inner realm that prepares you for that moment. Isn't that what you taught for years, Howard? The black church strengthened us, kept us sane, preached us toward one another, and reminded us that we were whole—so that when the moment came when we were required to stand up, we had the legs to do it; when we had to sit, we had the spiritual center to remain unmoved and unafraid.

Thurman The resistance that begins with "no" is more than the private concern of one group of people. The best wisdom that I can offer to this generation is to remember that liberation is not the exclusive concern of individual groups that can be contorted to fit prescribed needs and agendas. Liberation is inclusive and difficult. It requires that we expand our specialized needs and desires to include contiguous claims. The differences among race, class, gender, and sexual orientation liberation struggles exist, but only with regard to self-identification, methods of resistance, and historical journey. But the desire for liberation is universal and cosmological. When the trappings of identity politics are stripped away, we all want to be free.

Parks And I would add to that statement a reminder that freedom isn't free. Yes, I know that this is a cliché, but being free implies a responsibility to be equally proactive with regard to the freedom of others. Freedom does not do well in a vacuum. The more freedom there is, the better freedom thrives. The liberation of gays and lesbians does not impinge on the freedom of African Americans; it affirms freedom as the natural order of the universe.

Thurman You said that well, Rosa. Liberation extends beyond interest groups and national boundaries. The intensity of our struggles for liberation have exhausted not just our generation but also the generations that followed. We have pulled the world in around our ears and created a blanket to warm our own bodies while others suffer. Oppression anywhere is oppression in every cell of my body. Every holocaust implodes in my spirit. How do I enjoy a slice of pizza while Rwandans die? Liberation is an exercise in connection. My concern is that the circle of resistance has become small and exclusive. The boundaries of participation are being drawn according to the "ism" discomforts. Yet, liberation is an open project. It puts all those who struggle in the same battle on the same side.

Parks What the next generation needs to understand is that liberation is not just for the people; it is also for God. It is the story not just of the liberation of a people from the domination of low self-esteem and rabidly racist and usually invisible systems but also of the freeing of God from white supremacist ideologies and religious expectations and scholarship.

Thurman That's right, Rosa. When James Cone began his groundbreaking work, God was still God but was limited, in theological circles at least, to the epistemological cages of white theology. The question about God's freedom as an immutable attribute of divine character is an ongoing question that I will not debate; rather, I am concerned about God's functional freedom within the human-divine relationship. God is free as a matter of God's eternal essence and self-generativity, but not free because we are free to place limits on the relationship that may ultimately affect God's interaction in our lives and in the world.

In order to free God from racism's chains, Cone had to declare that God was black. In a battle of idolatries, black and white gods (refusing to allow yellow and brown gods to join the fight) slugged it out. It didn't take long for the idolatrous depiction of God as a supersonic version of an anatomically correct white male to shatter into jagged pieces. Each piece holographically reflected the God we had hoped for, who was a mirror image of us. And it was this breaking that opened the possibilities for God to be reflected in the images of people of other ethnic and racial origins and genders. But the images of God as red, black, yellow, and brown replicated the original fallacy.

No longer an imperial white male, God became the essence of men of every hue. My use of the word *men* is intentional, so let me be clear: God was transformed from a straight white male to a culturally relevant male, which was still problematic on several levels.

Parks I can remember marveling with other women that feminine images for God and the Holy Spirit have been lost in the mix.

My mind was racing at this point. I thought to myself, "Faith communities conflate the embodiment of Jesus and the Spirit of God in order to convert God into an anatomically correct male image that they can worship with fervor. In some circles, this misguided game is called 'pin the penis on God.'" Both Parks and Thurman looked at me and laughed. I had forgotten that they could hear what I thought.

Thurman Well, Sister, I wouldn't have put it quite like that, but I do agree that either all of humankind bears the mark of the divine or none of us does. We certainly can't be free if God is caged by our seething biases, because our freedom is in God. Accordingly, the potential for social liberation begins when the God who has always been free is freed from our human arrogance and ridiculous presumptions of control.

Writer Is this at all cosmological?

Thurman Absolutely. When we don't know or understand, we make things up out of whole cloth. We create a faith that will fit between our ears, something that will not challenge us or remind us of our mortality. Scientists believe that all of life originates from a black hole in the universe, but even if you prefer the biblical account (and I don't believe that the accounts of science and religion are antithetical), it is always darkness that is the site of creative genesis and yet darkness unnerves us.

We believe in the light and fear the dark, and in so doing we reject the comfort of "luminous darkness." I speak of a darkness described as an alternating reality and the dwelling place of God. I describe this darkness as a shimmering and infused reality that embraces and enfolds but does not foreclose the potential for the same goodness that we have ascribed to the light.

Parks Your discussion of "luminous darkness" was prophetic. Now we know that more than 75 percent of the universe is made up of dark energy that is propelling the planets away from one another at startling speeds. Dark matter is the unknown factor that may hold the cosmos together. No one knows for certain, but it is clear that darkness, whether luminous or not, will decide the fate of the universe. Dark energy will either pull the universe apart, or the mysteries of the unknown will hold it together. We can pretend that our protests and justice struggles are the end of the discussion, but the universe tells us something more.

Thurman What an interesting turn of events. The dark energy of dark people has been exploited and rejected around the globe, and now we find out that dark energy in the universe may determine the fate of the human family. One of my students described a deep-sea dive and his encounter with "the luminous quality of the darkness." He reported that immersion into this reality allayed his fears, improved his confidence, and gave him "peculiar vision."[4] This next generation will need courage and a peculiar vision to see beyond the apparent toward the possible. We are born, we live, and we die with a whole lot of life in between, but we cannot thrive unless the brackets of our physical existence loosen enough to allow the infusion of God's presence.

Parks Howard, you are speaking of liberation as more than the act of being delivered from specific oppressions, as the embrace of openness, fluidity, and un-power that accompanies a transformed consciousness. When we finally realized that we could not outfight or outnumber oppressors, we embraced the lack of power as the most powerful option of all. It is a bit like what you have been saying about dark matter, dark energy. It is unseen power that can affect lives on a profound level, Now, this is liberation in its finest configurations, emerging from the congruence and energy of a transformed consciousness.

"And," I thought to myself, "a transformed consciousness is the precursor of hope and potential." It wasn't long before a silence descended, dark and luminous. I plunged into its steadfast peace and slept.

Summary: Is There a Spiritual Aspect to Freedom?
Writer There are few successful freedom movements that proceed along rational and objective lines. From the fall of the Berlin Wall to the collapse of apartheid in South Africa, there is always an element that cannot be

quantified. Sometimes there is a leader like Gandhi or King who embodies this uncharted movement toward freedom, and sometimes all elements come together at the right time and the world changes suddenly.

Sarah I think that Rosa and Howard are trying to encourage this generation to pay as much attention to the leading of the Holy Spirit and the intentions of the human spirit as we have paid to the external manifestations of well-being. The freedom that is being sought may be available right now, but if the spiritual inclinations of its seekers are limited to what has been done and what is already known, or to what seems safe, then the chance of receiving and believing in a new way of being in the world diminishes drastically.

Writer So, this generation and those to come must determine what freedom means if materialism and national priorities do not set the definitions. Perhaps we will have to rethink the meaning of liberty and happiness.

Sarah You will also have to determine the scope of freedom and how you create supportive alliances even when liberation initiatives take the shape of a community that you don't support.

Writer We are talking about the transformed consciousness of the people as a prerequisite to a lasting liberation.

The Exercises

1. Choose one of the freedom movements that emerged during the twentieth century. Can you identify either contemplative or mystical elements related to the struggle for justice?

2. Examine the idea of "freedom" within your own personal context. Can you depict what it means to you using the artistic medium of your choice? Do not rely on public, national, or social images. Let this exercise reflect your inner images of freedom.

3. Choose a current freedom movement. Name three issues that make you most uncomfortable about their claims for liberation, and name three issues that you agree with. Explain why for each issue.

4. What is your personal experience of liberation? In one paragraph, describe a moment when you felt the most free; in another paragraph, describe a moment when you felt the most constrained. Are those moments tied to historical realities—for example, to slavery, immigration, or poverty? If not, explain, if you can, the source of your perception.

For Further Reflection

Rosa Parks
Brinkley, Douglas. *Rosa Parks*. New York: Viking, 2000.
Grant, Callie Smith. *Free Indeed: African-American Christians and the Struggle for Equality*. Uhrichsville, Ohio: Barbour, 2003.
Greenfield, Eloise. *Rosa Parks*. New York: Crowell, 1973.
Kohl, Herbert. *She Would Not Be Moved: How We Tell the Story of Rosa Parks and the Montgomery Bus Boycott*. New York: New Press, 2005.
Parks, Rosa, with Jim Haskins. *I Am Rosa Parks*. New York: Dial Books for Young Readers, 1997.
Parks, Rosa, with Jim Haskins. *Rosa Parks: My Story*. New York: Dial, 1992.
Parks, Rosa, with Gregory J. Reed. *Dear Mrs. Parks: A Dialogue with Today's Youth*. New York: Lee & Low, 1996.
———. *Quiet Strength: The Faith, the Hope, and the Heart of a Woman Who Changed a Nation*. Grand Rapids, Mich.: Zondervan, 1994.
Roop, Peter, and Connie Roop. *Take a Stand, Rosa Parks*. New York: Scholastic, 2005.

Periodicals
Black Enterprise, February 1993, 24.
Detroit Free Press, October 14, 2004.
Ebony, August 1971, September 1977, February 1988, January 2006.
Entertainment Weekly, December 19, 2003.
Essence, May 1985.
Jet, March 5, 1990; September 19, 1994, 22; September 23, 1996, 4; July 5, 1999, 32; December 13, 1999, 4; December 18, 2000, 8; September 18, 2000, 24; December 17, 2001, 10; February 25, 2002, 58; May 2, 2005, 17; November 21, 2005, 6.
Ms., August 1974.
Newsweek, November 12, 1979; December 26, 2005, 122.
New York Times, October 25, 2005, A1; October 26, 2005, A20.
PR Newswire, June 30, 1999; September 19, 2000; April 16, 2001; October 26, 2001.
Publishers Weekly, January 20, 1997, 402.
Smithsonian, December 2005, 34.
Vanity Fair, January 2006, 89.

Online
Troy State University Rosa Parks Library and Museum, http://montgomery.troy.edu/rosaparks/museum/.

Howard Thurman

Farmer, James. *Lay Bare the Heart*. New York: Arbor House, 1985.
Fluker, Walter Earl and Catherine Tumber, eds. *A Strange Freedom: The Best of Howard Thurman on Religious Experience and Public Life*. Boston: Beacon, 1998.
MacKechnie, George K. *Howard Thurman: His Enduring Dream*. Boston: Boston University, 1988.
Miller, Keith D. *Voice of Deliverance: The Language of Martin Luther King, Jr., and Its Sources*. New York: Free Press, 1992.
Thurman, Howard. *The Centering Moment*. 1969. Reprint, Richmond, Ind.: Friends United Press, 1984.
———. *The Creative Encounter*. 1953. Reprint, Richmond, Ind.: Friends United Press, 1972.
———. *Deep River: An Interpretation of Negro Spirituals*. 1945. Rev. ed., New York: Harper, 1955.
———. *Disciplines of the Spirit*. New York: Harper, 1963.
———. *Footprints of a Dream: The Story of the Church for the Fellowship of All Peoples*. New York: Harper, 1959.
———. *For the Inward Journey: The Writings of Howard Thurman*. Selected by Anne Spencer Thurman. New York: Harcourt, 1984.
———. *The Inward Journey*. 1961. Reprint, Richmond, Ind.: Friends United Press, 1973.
———. *The Luminous Darkness*. New York: Harper, 1965.
———. *Jesus and the Disinherited*. 1949. Reprint, Richmond, Ind.: Friends United Press, 1981.
———. *Meditations of the Heart*. 1953. Reprint, Richmond, Ind.: Friends United Press, 1972.
———. *Mysticism and the Experience of Love*. Wallingford, Pa.: Pendle Hill, 1961.
———. *The Negro Spiritual Speaks of Life and Death*. New York: Harper, 1947.
Meditations for Apostles of Sensitiveness. Mills College, Calif.: Eucalyptus, 1948.
———. *The Search for Common Ground: An Inquiry into the Basis of Man's Experience of Community*. 1971. Reprint, Richmond, Ind.: Friends United Press, 1986.
———. *With Head and Heart: The Autobiography of Howard Thurman*. New York: Harcourt, 1980.

Periodicals

Atlantic Monthly, October 1953.
Black Enterprise, July 1981.
Life, April 6, 1955.
Nation, January 5–12, 1980.
New York Times, March 22, 1953.

At the end of the day, I am not so much concerned about my own personal influence as I am concerned with carrying out a legacy and doing my best to guarantee that that legacy lives on with future generations. I see myself, along with many other people, as a part of a tradition of struggle, and this is what I try to convey to students, workers and prisoners.
—Angela Davis

5. A Revolutionary Liberation: Freedom and Wholeness

There are some African-Americans who will resist the idea of attempting to create a new vision and will insist that either nationalism or integration will suffice. . . . They will appeal to either Malcolm or Martin, or both, as if we were still living in the 1960s. But history does not wait for those who are spellbound by romantic visions of past heroes. If we study our history critically, we shall know that a true understanding of Martin and Malcolm pushes us beyond them both to the creation of a future for which they died but which they saw only imperfectly.
—James Cone

In this conversation, Malcolm X and Harriet Tubman discuss their different approaches to revolutionary action, and their historical struggles to lead the community toward a tangible freedom. The liberation that they describe requires not just a declaration of release from bondage but also the freedom to be fully human in a practical, spiritual, and cosmological sense.

The Elders

Harriet Tubman, 1819–1913[1]

I had reasoned this out in my mind; there was one of two things I had a right to, liberty, or death; if I could not have one, I would have the other; for no man should take me alive; I should fight for my liberty as long as my strength lasted, and when the time came for me to go, the Lord would let them take me.[2]

Harriet Tubman (born Araminta Ross) was the daughter of slaves held in Maryland.[3] Her childhood is a patchwork of various scenarios of abuse at the hands of slaveholders. Biographies recount regular beatings that culminated in a severe blow to the head for refusing to help tie up a slave who had attempted to escape. She was on a slave plantation in Maryland when the story began to circulate that Harriet would be sold with her brothers to a chain gang in the deep South. It was at this point that her lifelong desire to escape became a concrete plan. She escaped to Philadelphia and made many trips to Canada. She returned to the South more than twenty times to rescue family members and others.

Harriet supported many causes. She raised funds for John Brown's rebellion, cared for Union soldiers during the Civil War, participated in the suffragist movement, attended the first convention of the National Association of Colored Women, and worked on issues related to voting and education for freed slaves. She left several acres of land and a home for the aged to the African Methodist Episcopal Church. She is known as the Moses of her people because she led so many to freedom.

Malcolm X (El-Hajj Malik El-Shabazz), 1925–1965

If you are not ready to die for it, take the word freedom *out of your vocabulary.*[4]

Malcolm X was born Malcolm Little in a midwestern town in Nebraska. He was one of eight children, the son of an activist Baptist preacher and a stay-at-home mom. Because of his father's liberationist activities, the family was subjected to death threats, the burning of the family home, and continued harassment by white supremacist hate groups. The harassment culminated in the murder of his father, Earl Little. The case was never solved. His mother, Louise, was committed to a mental institution after the murder of her husband, and the children were raised in foster homes. Malcolm drifted into a life on the margins where crime, drugs, and prostitution inevitably led him to prison. It was during his period of incarceration that he advanced his education and become a convert to Elijah Muhammad's Nation of Islam.

It was not long before Malcolm was nationally known for his radical oratory and his leadership position with the Nation. But leadership with the Nation was only one aspect of his journey. Malcolm became dissatisfied with the disparity between Elijah Muhammad's public and private values. When he left the Nation, he turned toward Islam, made his first pilgrimage to Mecca, and began to explore human relationships beyond race and ethnicity. When he was killed on February 21, 1965, by three gunmen in the Audubon Ballroom in New York City, he was moving toward an egalitarian view of liberation.

The Conversation

I couldn't help a bit of excitement as I stepped out into what seemed to be a front porch. The setting was quite a change from the meeting room atmosphere, but in truth nothing was as it seemed. Although I didn't know where these interviews were taking place, I always had the impression of place and time. Tonight I could "see" evening stars sprayed against the thick darkness like a brilliant fireworks display. I tried to find the North Star with no luck, wondering what it must have been like to escape from bondage with eyes fixed on the heavens for direction and inspiration. The story of escapes north suggest that fugitive slaves used the North Star to map their way to freedom. I imagine a different scenario.

The universe reminds us of our great potential and origins every time we look up or turn inward to reflect. When escaping slaves gazed at the heavens to get their northbound bearings, they were also reminded that they were connected to powerful cosmological realities that dwarfed the limited options of the slave owners. The entire cosmos testifies to the predominance of creativity and freedom. Now that's something to ponder during a long walk. Perhaps the most revolutionary acts arise when we experience a reorientation from individual limits to spiritual interconnections, when we recognize the differences between the power of violence and physical force and the power of a moral order.

This would be a particularly meaningful meeting for me because it evoked memories of my own past. The fortieth anniversary of the Black Panther movement was celebrated recently. During the 1960s, they met in New Haven, Connecticut, my hometown. I grew up in Yale's backyard during an era when the campus was a hotbed of protest and planning. While I was old enough to know what was going on, I was not old enough to have the agency to make my own decisions as to how I would involve myself. My father, a quiet but wise man, said no to attendance at Black Panther planning meetings (I was too young to be out that late) and no to a small gathering on Dixwell Avenue in the heart of New Haven's black community to hear the new Black Muslim leader Malcolm X . (My father said, "I don't know where he stands on Jesus, and we are Christians; we've already made that choice.")

His yes came with regard to our participation in the freedom rides and the support of Martin Luther King Jr. My father agreed with King that revolutions need not be dramatic or violent; rather, persistence, focus, and the in-breaking of a liberating God would win the day. This was an era when activism was the ordinary business of thoughtful people. but it was also a time of great uncertainty. No one knew what would happen when the nonviolence of Martin Luther King Jr. hit the wall of American violence. The answer came quickly with the deaths of the children in a Birmingham church, the deaths of Emmett Till, Medgar Evers, and Viola Liuzzo, to name a few.

People began to wonder if King's strategies could work in the face of radical evil. This was not a personal rejection of King, just the certain belief that such resolute violence against unarmed people should be met with a different type of revolutionary action. So on this day, I certainly wanted to talk with Malcolm X. When my reverie ended, Tubman was already rocking on the porch. She wore a simple but pretty black dress that looked like it would travel well. I guess I expected to see the long outfit and head wrap that became her trademark image for all who studied her story in school.

Tubman [embracing me] Heavens, Child, can't a woman update her wardrobe? I wish I'd had wash-and-wear on the trail to Canada.

I didn't see Malcolm yet, so I embraced her and dove right in to questions about the trail to Canada. Her laugh was infectious but a little too hearty for one who had spent her early years hiding from slave catchers and their dogs.

Tubman No, let's wait for Malik. He'll be here in a minute.

It wasn't long before El-Hajj Malik El-Shabazz could be seen coming down the road that seemed to end at this porch. His attire made no sense to me either. First, I had to deal with Tubman in the proverbial basic black dress. Now Malcolm in jeans? I remember nothing but business suits, ties, and hats.

Malcolm [smiling as he shook my hand] You wear your position if you are smart. Mine was that we are human and we are respectable men and women. Since respectability in the mind of the dominant culture at the time meant ties and so forth for men and dresses and gloves for women, the message and the clothes had to match. But enough about me. This is important work, so let's get started.

Writer Ah Sir, what shall I call you?

Malcolm Well, certainly not "Sir." Use my slave name. It helps those who most need to hear the message to know who you are talking about, but be sure to note that my name for all time is El-Hajj Malik El-Shabazz.

I could tell that it was going to be hard to stick to topic with these two, because the peripheral issues were just as interesting as the topic of revolution.

Tubman You haven't said anything out loud, but I can hear you thinking. I know it's difficult to figure out where to start, but as I used to tell my passengers on the underground railroad, start right where you are and then go somewhere else.

Writer Did you really say that? I've never heard that quote before.

Tubman Well, we didn't have tape recorders, so you only have a record of what I remembered to repeat. But there was so much said, so much going on.

It was clear that I was going to have to work at not being a groupie with these two.

Writer I guess I don't really have any questions. Just talk to each other about revolution within the context of liberation and what it means to you.

Malcolm I hope that you don't think that we have a neat revolutionary package for you, because the very meaning of revolution requires the creative guerrilla thinking that precedes any action. In my day, any revolutionary worth his or her salt was a thinker, a strategist, and a spiritually open vessel. They look at my speeches now and talk about the brilliance, but I always knew that the words were coming *through* me, not *from* me.

Tubman I know what you mean, Brother. The act of revolution is to take a step in a direction that you've never been in before. The task is to walk right into the face of danger, risking everything knowing that you don't really have anything at risk. If your life is not your own but is poured out for others, what do you really have on the table? I used to say that the only one gambling was God because, given free will, God never knew whether we would walk by faith or not.

Malcolm Good point, Sis. The greatest risk is not living up to the tasks that the universe lays at your feet. You have to decide to accept your life mission. As you walk, it unfolds. And, it's not that you are without fear; it's that fear and insecurity can't stop the show. It can buy a ticket and watch, but it can't stop the fulfillment of your moral responsibility unless you relinquish your power to act.

Tubman Revolution for me meant seeing a social and political system that was broken and knowing that even if I did not have the power to change the system, I could put a spoke in the wheel of oppression. Does that language sound familiar to you? Well, I was talking to Bonhoeffer the other day, and I picked that up from him. My revolution was to confront the system not with battle but with a decision.

I had no army, no power to resist the villainy of a nation, but I could make the decision that I would not stay. I chose to use what I had, my feet, and to encourage others to use theirs. Either there is a God or there is not. We were either going to be caught and killed or we were going to be free. Each time I was not caught or killed, I had to place the same cards on the table again. When God shows up, you have to decide that God can show up again.

Malcolm You don't always win that lottery. Sometimes God allows martyrdom, but if our lives are connected to every other life, and if we are given a designated period of time to serve others, why not take the risk? You have to die of something, and this was a brilliant something. I personally know what MLK experienced just before his mountaintop speech. I was also in that place of profound peace just before my assassination. It was a precursor of where I am now. The peace is so profound that even the foreknowledge of your own demise holds less interest than your willingness to follow the path set before you. When the bullets hit that type of resolve, your life, the life that they think that they have taken, explodes throughout history so that others will know for eons to come how to live and die with purpose and an ethics of responsibility.

Tubman I agree that for some causes you have to break the egg to make the omelet. You and Martin were larger than life before and after your assassinations, but in other cases like mine, it is very different. If I had been killed on the trail, I doubt that my life would have exploded into shards of freedom significant enough to pierce the heart of oppression. Instead, my death during an effort to escape would have been offered as a cautionary tale that escape was fruitless and that, as a people, we had to live within the assigned constraints.

Each of us will cross over from earth life to spirit life. We all eventually die, but why hand the reasons and circumstances of our deaths over to the ones who have taken away our options for life? Dying is fine, but dying without choices is tough. I decided that if I was going to die, I would rather do it on a walk toward freedom.

Malcolm Can I say a bit about personal revolution before we go on, because revolution is an internal as well as an external process? While the revolution was forming in the culture, it was stirring in me. I was becoming and shedding and growing and learning. The revolution has to happen within before a leader can model publicly that same spiritual inclination before the people. What seems to be needed in the twenty-first century is the commitment to personal revolution. I guess I'm referring to a revolution of values.

If the values that have been inherited no longer have meaning to the present generation, then the task is to identify the contexts, norms, and modalities of moral flourishing that *will* be meaningful.

Tubman But why try to carve out new value systems, new pleas for the government to fulfill its moral obligations to the people? Why stay in North America at all? Is this country the only place where Diasporan people can find solace? This generation has more access to travel than ever before, and yet it seems stuck. They seem unhappy with their plight but unable to escape the expectations of previous generations. After the Katrina catastrophe, the anger from most people of color was intense, but everyone returned to business as usual. There will be other Katrinas, if history is any predictor of the future, so shall we continue to pretend that this is the promised land?

It is almost as if the boundaries of the plantation have expanded. Do they have this generation penned in with electronic devices? If not, and if the system that seemed to offer opportunity and a dream of diversity no longer works, why do they stay?

Malcolm I guess it is familiarity that keeps them there. There will be those who say that the United States is the greatest country in the world. But that depends on the measurement criteria. The question is not one of patriotism or loyalty; the question is whether or not this social experiment works for some, all, or none of the people. We are asking an unjust system to produce a gift-wrapped version of freedom, to set the captives free. Although freedom as an in-breaking event has happened in history, it hasn't happened in the United States. Let me just repeat what I said a while back.

> It is impossible for a chicken to produce a duck egg. Even though they both belong to the same family of fowl. A chicken just doesn't have it within its system to produce a duck egg. . . . The system in this country cannot produce freedom for an Afro American. . . . It is impossible for this system . . . to produce freedom right now for the black person in this country. And if a chicken did produce a duck egg, I'm quite sure you would say it was certainly a revolutionary chicken.[5]

I'm not certain that freedom can be birthed or sustained on this blood-soaked soil. If not, then there is the responsibility to explore options. If this nation is the promised land, then grow crops and worship your God; if not, then plan a pilgrimage, even if you have to walk like my sister.

Tubman And the way that you determine whether or not something is working for you is if it benefits the least. You may be doing fine, but if others are being treated as less than human, and if walls are being put up with your tax dollars to keep them out except to toil in the factories, then it may be working for you now, but the poison that infects them will spread. Then you use the best tool of a revolutionary, your voice. They say that I kept folks on the trail at gunpoint, but actually it was the stories that we told at night about Canada and what it was like to live in freedom that kept folks on the trail.

It was an open night sky that testified to myriad options and the power of its creator that inspired the will to keep walking. Certainly, I would have shot a deserter to save a group, but you can't force liberation on anyone.

Malcolm Revolution does not come because one army fights another. It is the articulation of a changed perspective that undermines the very foundations of the old. In an interesting way, the song "The Revolution Will Not Be Televised" was right. We created worlds with our words and actually reshaped the public image of dark people in ways that matched inner realities.

We articulated what the people thought and confronted the fear that kept them from speaking. I always spoke as one coming from a community that shared beliefs and commonly held experiential wisdom. I really didn't say anything new—I just said it out loud and broke through the fact-based fear among African Americans that righteous speech would result in immediate death. The announcement was to the dominant culture that a revolution had taken place and not a shot had been fired. This was a quiet transition from one state of being and mind to another. It had happened right under the nose of those who supported domination, and paradoxically, it was an unexpected benefit of segregation. Walling off the community from public life fostered spiritual growth and communal interconnections.

My task was to announce the revolution to the nation and at the same time to say to the African American community, it's go time! The community had a long tradition of multilayered discourse, using one narrative to secretly transmit another message. Slave songs held information about escape. Discourse was opaque with hidden meaning, and we were still doing that in the sixties when I began the revolutionary act of saying what we thought directly. Even my beginning speeches that focused on the "white devil" offered a cathartic outlet for the anger that was continually suppressed. Whether they were in the Nation of Islam or not, they were angry. They still are, but there is no one to blame.

My advice to those now struggling with the threat and promise of revolutionary action is that you are not required to make a choice between violence or peaceful demonstration. That day has passed. Now, effective revolutions transform consciousness and spirit.

Tubman Well, that's all well and good, but the threat of violence that you brought to the table during the civil rights era was a necessary adjustment to the doctrine of love and peaceful resistance. Every peaceful step was met with violence. It became apparent that nonviolence alone, without the threat of militant defense, would falter. The best example I can think of was the first march from Memphis to Jackson that James Meredith attempted. It ended on the second day when he was ambushed by Aubrey James Norvell.[6] When Martin Luther King, Stokely Carmichael, and others attempted to complete the march, they were shoved into the dirt by Mississippi state troopers.[7]

The mere juxtaposition of the word *power* with the word *black* was enough to frame an era of activism and violence that seemed to be an incongruous ending to the Civil Rights movement. In retrospect, it seems a rational response to the institutionalization of white supremacy and the escalating violence directed at peaceful protestors. Many of the liberation movements presumed that the power of self-actualization would be automatically ceded as a reward for generations of uncompensated labor and abuse. One of my contemporaries, Frederick Douglass, thought otherwise. He said:

If there is no struggle there is no progress. Those who profess to favor freedom and yet depreciate agitation, are men [sic] who want crops without plowing up the ground, they want rain without thunder and lightning. They want the ocean without the awful roar of its many waters.

This struggle may be a moral one, or it may be a physical one, and it may be both moral and physical, but it must be a struggle. Power concedes nothing without a demand. It never did and it never will. Find out just what any people will quietly submit to and you have found out the exact measure of injustice and wrong which will be imposed upon them, and these will continue till they are resisted with either words or blows, or with both. The limits of tyrants are prescribed by the endurance of those whom they oppress. In light of these ideas, Negroes will be hunted at the North, and held and flogged at the South so long as they submit to these devilish outrages, and make no resistance either moral or physical. . . . If we ever get free from the oppressions and wrongs heaped upon us, we must pay for their removal. We must do this by labor, by suffering, by sacrifice, and if needs be, by our lives and the lives of others.[8]

Malcolm Douglass is right. The peaceful revolution that we all wanted occurred when the endurance for abuse ended. Both Martin and I used rhetoric as a weapon of choice to empower the people and to warn the dominant culture of the changing order. We never say so, but the revolution was successful in many ways. Liberation came but was not completed. It did not fail; rather, it stalled. African Americans reclaimed their inner power and demonstrated through marches and the rhetoric of liberation a determination to be free.

With each liberation comes exposure to new possibilities for bondage. The new task is to revolutionize the community to ensure its safety and well-being. Integration never was the way and is not now the hope of the future. Stokely Carmichael was right when he described integration as "an insidious subterfuge for white supremacy."[9] Those who want to continue that struggle are engaged in a futile enterprise that forces them to choose a synthesis of cultures and values that rarely succeeds.

The next phase of revolution will come with the reclamation of entrepreneurial instincts, individual and communal health, and spiritual focus. We should teach our children because we love them and the system seldom does. We must align with Africa because she is our mother. We can't fight brothers and sisters because they have been declared enemies by a society that flourishes amid the false dichotomies of good and evil. Now that's revolution.

Do you remember the words of Stokely Carmichael uttered during his Berkeley California speech in October 1966?

> We must question the values of this society, and I maintain that black people are the best people to do that since we have been excluded from that society. We ought to think whether or not we want to become a part of that society. . . .
> I do not want to be a part of the American pie. The American pie means raping South Africa, beating Vietnam, beating South America, raping the Philippines, raping every country you've been in. I don't want any of your blood money. We must question whether or not we want this country to continue being the wealthiest country in the world at the price of raping everybody else.[10]

A real revolutionary includes in specific struggles, the liberation of all people who are suffering.

Tubman I agree. Soon it will be time to determine whether this great multicultural experiment works on any real level. If oppression just changes its clothes and greets your children in a different costume, is that all right? If it rocks your grandchildren while humming lullabies of self-hatred and inferiority, should you stay? They used to call me the Moses of the Underground Railroad. Are y'all waiting for a Moses to lead you out when airlines fly daily to other places?

Can you imagine a life somewhere else with less in the way of material goods but more in the way of relationships and life purposes? If I were there now, I'd be flying instead of walking, but I'd be looking for a place of respite where my children would not find themselves in opposition to the entire world because of their consumption of most of the world's resources. I'd be trying to learn from Brazilian healers and Inuit shamans. I'd be visiting Africa and acquainting myself with the streets of Europe. The most revolutionary act in the twenty-first century is the globalization of your thinking. And this cannot be done from an armchair in front of the television.

Broaden the boundaries of your mind, follow the drinking gourd, seek neighborhoods beyond your tree-lined streets, and a revolution in child rearing and family relationships and community goals will inevitably follow. You can't even do church in the same way after you've experienced expressions of faith outside of the United States.

Malcolm Are you saying that power in the next century will require negotiation of claims and the common purposes of the oppressed?

Tubman Well, I guess that's a fancy way of putting it.

Malcolm The understanding of power and how it is used in the interest of justice is the entire agenda of the twenty-first century. Imagine if you will that there is a village of one hundred people and thirty-five live in one section with all of the services and goods, technology, and weaponry while the rest dig sweet potatoes and die of curable diseases. Imagine that the majority live in subsistence conditions while the thirty-five constantly diet because they eat so much. How long do you think that the thirty-five would be allowed to exploit the labor of the others for their benefit?

The two-thirds world represents the majority. The power to equalize and distribute goods and services is not far off. It requires a revolution in thinking not unlike the revolution that occurred when the phrase "black power" became the call for self-empowerment, and black theology became the first vehicle for imagining ourselves whole.

Tubman But isn't it even more than that? Aren't we talking about a revolution in black consciousness? Scholar James Cone says this about your early days, Malcolm, when you were challenging the powers of government:

> It is one thing to recognize that the gospel of Jesus demands justice in race relations and quite another to recognize that it demands that African Americans accept their blackness and reject its white distortions. When I turned to Malcolm, I discovered my blackness and realized that I could never be who I was called to be until I embraced my African heritage—completely and enthusiastically, Malcolm put the *black* in black theology.[11]

Writer Yes, but descriptions of blackness must allow shades of difference and variation to emerge. To use the phrase that Victor Anderson coined, "the blackness that whiteness created" tends to be a monolithic gloss that distorts and often obliterates the contributions of black women and pretends that everyone has the same racial and identity priorities. Anderson asks for more.

> To press beyond ontological blackness, African American theology needs a public theology that is informed by the enlightening and emancipatory aspects of postmodern African American cultural criticism.[12]

Anderson is suggesting that we come out of our private caves, emerge from the shadows where we live private lives, so that we can bring the cultural criticism of the kitchen table to the public square. We need to debate and discuss those issues that are corroding community life. If we are for freedom, is it freedom for all or freedom for some? Are we willing to protect the right to lifestyles that don't agree with our specific social choices, or must freedom take on the shape of our personal set of moral values?

Tubman This is not the way of the ancestors. Acceptance of difference was enfolded into everyday life. If you ask me, the liberation movement is stalled because too much was expected. It was supposed to even the playing field, resolve historical resentments, rebuild self-esteem, and leave everyone happier and more fulfilled. It is like whipping a horse because it can't run a race on ten different tracks at the same moment. More horses had to be let into the race so that the means of empowerment and the critical reclamation of history could reflect the varied realities of the black community. When we felt that we had to present a united front, we hid our differences. Now the black community needs to embrace the complex layers of reality that it inhabits.

Writer What would you say if I suggested that freedom is like a quantum particle, that it takes on the meaning that you assign to it?

Malcolm I'd say that you have far too much time on your hands. I just knew that you'd get science into the conversation. The word is out.

Writer No one can determine what quanta are; they are both wave and particle. They resemble matter only when we fix our gaze on them; otherwise, they are something else entirely. And since we are all made up of quanta as the smallest unit of reality that we can grasp, we are also all things and no-thing. We have potential that extends beyond our comprehension but can circumscribe our reach in shockingly effective ways.

Malcolm Deep! I was wondering how cosmology would fit into our discussion about justice and the liberation of people of color around the world. But I'm not interested in trendy quantum connections. I want to be certain that no matter how the world is configured, the new frontiers of knowledge will guide and bless the people.

Tubman As I traveled back and forth from Canada and back to the South, I felt a sense of knowing—that I realized early on didn't come from my little ol' head. There was more. I would look up at the North Star, and it was as if the map that led from despair to hope was threaded throughout the night sky. The old folks used to call it intuition and instinct, but the truth of the matter is that we only know in part. When we connect to some greater wisdom, we can make our way, even if we don't understand the path.

Malcolm Science backs you up, my sister. Just suppose that we encode our freedom in our bodies, so that even when we feel most separate and alone, we are carrying wholeness inside. Now, I'm no scientist, but I read a lot, and this idea of wholeness encapsulated in each piece of matter is intriguing. They call this concept holography. If the writer would be so kind as to summarize holographic science for us, then we can see if this connection between cosmology and liberation has legs.

Writer Who, me? Well . . . I know that the idea of holography is included in David Bohm's view of the universe, but I'll let Michael Talbot describe precisely what a hologram is.

> In the movie *Star Wars*, Luke Skywalker's adventure begins when a beam of light shoots out of the robot Artoo Detoo and projects a miniature three-dimensional image of Princess Leia. Luke watches spell bound as the ghostly sculpture of light begs for someone named Obi-wan Kenobi to come to her assistance. The image is a hologram, a three-dimensional picture made with the aid of a laser. . . . What is even more astounding is that some scientists are beginning to believe the universe itself is a kind of giant hologram, a splendidly detailed illusion. . . . There is evidence to suggest that our world and everything in it—from snowflakes to maple trees to falling stars and spinning electrons—are also only ghostly images, projections from a level of reality so beyond our own it is literally beyond both space and time.[13]

Tubman I don't know if I can buy into the idea that the entire universe is a projection or an image. Are you saying that slavery and oppression are illusions? I know that every step that I took toward Canada was real, and I had the blisters to prove it.

Malcolm I don't think that's what Talbot means. I for one am grateful that science tells us that what we see is not what we get, and that what we know is just a scintilla of what is actually available to us. Now that is a liberating thought.

Writer If I may, it is not that everything is fake; it is that everything is more real than we could have imagined. There is, according to one of Einstein's protégés, David Bohm, an order (implicate) that we cannot see and an order (explicate) that we can perceive. We evolve from wholeness. Now, whether that wholeness is actual or spiritual, or whether there is a difference between the two, I can't say.

Tubman So if one person has knowledge, all have it?

Malcolm Yes, and just like a hologram, we each carry wholeness in our bodies and spirits. We may think that we have only a piece of the story, but if holographic science is correct, we carry the entire story of our people in our individual lives. No amount of dissection will disclose the location of this wholeness; it is just there as a gift that we seldom take time to recognize or acknowledge. There is a scientific revolution going on, and folks don't seem to know it. This revolution of perspective will ultimately turn out to be a revolution of spirit and mind.

Tubman There is an often-told story of a zoo-born lion who cringes when he sees a photograph of a Masai warrior. It is well known in Masai country that lions fear the tribe members and will not attack a lone warrior tending hundreds of cattle. But what would a zoo-born lion know of this unless the entire story is encoded in every lion's DNA? OK, that has to be ridiculous, but could there be a level of consciousness that is shared in such a way that species information is available even if the information is of no use in a city zoo? Now that's enough to get you out of your comfort zones!

I don't know what it means, but when we began this conversation, I was advocating leaving a nation that refused to offer the most basic ideals of liberation. Now I'm beginning to think that we all carry our liberation within us. Perhaps that is the message that we were supposed to bring. Trust the story within you, and live into it with courage and faith.

Malcolm The journey is from comfort zones to the risky places where neighbors are in need of help and God is waiting to point the way toward the future. My advice to this future generation would be: prepare yourselves to lead. We are not called just to follow; we are also called to lead. Leadership often requires that you walk where others will not and speak the truth that others know but will not say. Real revolutionary acts are seldom violent. It is an act of revolution to understand that the story within humankind is the story of God and the story of the universe. There is only one story. As-salaam-alaikum.

Tubman Wa-alaikum-as-salaam. I offer this to the next generation: may the peace of God go before you, cover your scent so that the dogs won't find you, and strengthen your heart for the journey. And if you lose your way, lift your head toward the heavens and follow the drinking gourd.

Summary: Do Liberation Initiatives Require a Revolution of Values?

Writer I thought I heard Malcolm saying that the most powerful revolutions come in discourse and reorientations of our social and personal commitments.

Sarah Yes. Long before Tubman starts to walk toward the North, she has a revolution within her spirit and decides that the common wisdom that advised safety and survival was not for her.

Writer I hear their call to recognize the role that individuals play in the overall movement toward liberation. Martin Luther King needs the challenge of Malcolm to keep the momentum toward legal and institutional change. Malcolm hears Martin's use of "love" as a radical option for revolution, and then discovers the transnational "neighbor" connections at Mecca.

Sarah They were blessed to be able to lead, if only for a short time. The places of greatest blessing are also the places of greatest risk. We have to decide what we are willing to risk and when.

Writer They were both led by something greater than themselves. Tubman looks to the skies for guidance and then starts walking, Malcolm translates his anger and his spiritual transformation into "accessible resources for the community."[14] Tubman had to divest herself of values instilled during slavery; Malcolm had to divest himself of the residual resentment resulting from daily life in a racially divided nation.

The Exercises

1. What tasks has the universe laid at your feet? Do you feel a call or spiritual inclination toward participation or leadership of a cause that might lead to social change? Describe one personal revolution that comes from within you but also seems connected to a larger purpose.

2. Can you identify three peaceful social revolutions that are pending in the twenty-first century?

3. Imagine that you are either Tubman or Malcolm. Write a three-page story. If you choose Tubman, imagine that you are leading people out of bondage today. Describe the bondage. What are they leaving and where are they heading? If you choose Malcolm, imagine that you are a public spokesperson "speaking truth to power." Name the current injustice (you may choose a national or international issue). What revolutionary solutions are you posing?

4. Consider an old conflict in the world that is not resolved. Propose a one-page outline for resolution of the issues. Think out of the box. Ordinary solutions have not worked. Using a cosmological and revolutionary perspective, suggest options.

For Further Reflection

Harriet Tubman

Blockson, Charles L. *The Underground Railroad*. New York: Prentice Hall, 1987.

Bradford, Sarah. *Harriet Tubman: The Moses of Her People*. New York: Corinth, 1961. (Reprint of second edition originally published in 1886. First edition published in 1868 was titled *Scenes in the Life of Harriet Tubman*.)

Clinton, Catherine. *Harriet Tubman: The Road to Freedom*. Boston: Little, Brown, 2004.

Commire, Anne, ed. *Women in World History: A Biographical Encyclopedia*. Waterford, Ct.: Yorkin, 1999–2000.

Humez, Jean. *Harriet Tubman: The Life and Life Stories*. Madison: University of Wisconsin Press, 2003.

Larson, Kate Clifford. *Bound for the Promised Land: Harriet Tubman, Portrait of an American Hero*. New York: Ballantine, 2004.

Schroeder, Alan. *Minty: A Story of Young Harriet Tubman*. New York: Dial Books for Young Readers, 1996.

Still, William. *Still's Underground Rail Road Records, Revised Edition, With a Life of the Author. Narrating the Hardships, Hairbreadth Escapes and Death Struggles of the Slaves in their Effort for Freedom*. Philadelphia: William Still, 1883.

Malcolm X (El-Hajj Malik El-Shabazz)

(Resources compiled by Dorothy Ann Washington, librarian, Black Cultural Center, Purdue University)

Asante, Molefi. "Malcolm X as Cultural Hero." In *Malcolm as Cultural Hero and Other Afrocentric Essays*. Trenton, N.J.: Africa World Press, 1993.

Breitman, George. *The Assassination of Malcolm X*. New York: Pathfinder, 1976.

Cone, James H. *Martin & Malcolm & America: A Dream or a Nightmare*. Maryknoll, N.Y.: Orbis, 1991.

———. *The Last Year of Malcolm X: The Evolution of a Revolutionary*. New York: Merit, 1967.

David, Donald. "Odd Passings and Other Assassinations." Ch. 5 of *Big Book of Conspiracies* by Doug Moench. New York: Paradox, 1995.

Davis, Thulani. *Malcolm X: The Great Photographs*. New York: Stewart, Tabori & Chang, 1993.

Johnson, Timothy V. *Malcolm X: A Comprehensive Annotated Bibliography*. New York: Garland, 1986.

Kazi-Ferrouillet, K. "Afrocentricity, Islam, and El Hajji Malik Shabazz (Malcolm X)." *Black Collegian* (January 1993): 144-47.

Kieh Jr., George Klay. "Malcolm X and Pan-Africanism." *Western Journal of Black Studies* 19, no. 4 (Winter 1995): 293-99.

Malcolm X: FBI Surveillance File. Wilmington, Del.: Scholarly Resources, 1978.

A Malcolm X Reader. New York: Carroll & Graf, 1994.

Sales, William W. *From Civil Rights to Black Liberation: Malcolm X and the Organization of Afro-American Unity*. Boston: South End, 1994.

"The Strange Death of Malcolm X." *Black Issues in Higher Education* (March 1993): 124-29.

Strickland, William. *Malcolm X: Make It Plain*. New York: Viking, 1994.

X, Malcolm. "Alex Haley Interviews Malcolm X." *Playboy Magazine* (May 1963).

———. *The Autobiography of Malcolm X*. New York: Ballantine, 1973.

———. *By Any Means Necessary*. Ed. George Breitman. 2nd ed. New York: Pathfinder, 1992.

———. *Malcolm X: Speeches at Harvard*. New York: Paragon House, 1991.

———. *Malcolm X: The Last Speeches*. New York: Pathfinder, 1989.

Race is for me a more onerous burden than [HIV]AIDS. My disease is the result of biological factors over which I have no control. Racism is entirely made by people, and therefore it hurts infinitely more.
—Arthur Ashe

6. Liberated Bodies, Liberated Lives: Embodying Freedom

How do we wean ourselves from being masters and owners? I think that mystical spirituality of oneness with nature is the best preparation for the other life we are looking for. Dealing sacramentally with bread and water, one's own body and our nonhuman sisters and brothers, and with energy, the cosmos itself will grow from the abyss that is our domination-free ground.
—Dorothee Soelle

This chapter considers issues of freedom in the body. Here, Audre Lorde, Fannie Lou Hamer, and George Washington Carver discuss the intersections of healing as a personal, political, and communal event. The chapter includes discussions of the politics of food, public service, and issues of gender and sexuality.

The Elders
Audre Lorde, 1934–1992

We who are Black are at an extraordinary point of choice within our lives. To refuse to participate in the shaping of our future is to give it up. Do not be misled into passivity either by false security (they don't

mean me) or by despair (there's nothing we can do). Each of us must find our work and do it. Militancy no longer means guns or high noon, if it ever did. It means actively working for change, sometimes in the absence of any surety that change is coming. It means doing the unromantic and tedious work necessary to forge meaningful coalitions. And it means recognizing which coalitions are possible, and which coalitions are not. It means knowing that coalitions, like unity, means the coming together of whole, self-actualized human beings, focused and believing, not fragmented automatons marching to a prescribed step. It means fighting despair.[1]
—Audre Lorde

Audre Lorde was an activist, artist/poet, and creative liberationist. As a black lesbian, she found herself at the intersection of multiple oppressions. She rallied women and created international alliances to challenge sexual discrimination. She started a feminist publishing company and was educated at Hunter College. Lorde used her national presence to initiate public conversations about the role of women and the accountability of the medical community to women's needs. She also challenged white women on the issues that divided their interests from the interests of women of color. Her passionate but loving critique is one that is still being engaged today. Her literary works are sometimes romantic but also daring and forthright. She viewed liberation as the option of all, and not some, of the people. Her essays consider the struggle for freedom to be an intrinsic part of human existence. In her self-described role as a "black, lesbian, feminist, mother, warrior, poet," Lorde found moments of liberation in everyday life, in her relationships with others, and even during her losing battle against cancer.

Fannie Lou Hamer, 1917–1977

Nobody's free until everybody's free.[2]

Fannie Lou Hamer was a hard-working woman whose life became an example of the power of ordinary people to change their lives. She was born in Mississippi during an era of violence and oppression. Freedom was no closer than heaven for those caught in the system of sharecropping. She dropped out of school, picked cotton, and married. This would have been the complete story of her life if she had not decided to change her fate by registering to vote.

Her ensuing arrest and harassment strengthened her resolve to put the same energy into her own liberation and the liberation of her people that she had expended on cotton crops. The consequence for this choice was a beating so horrific that one wonders how she survived. She continued to work for the Student Nonviolent Coordinating Committee registering voters, and during the 1964 Democratic Convention, she challenged the all-white delegation from Mississippi. When she died, she bore in her body scars that mapped a path toward freedom.

George Washington Carver, 1864–1943

When I was young, I said to God, God, tell me the mystery of the universe. But God answered, that knowledge is for me alone. So I said, God, tell me the mystery of the peanut. Then God said, well, George, that's more nearly your size.[3]

George Washington Carver was born to enslaved parents in Missouri. His father died before he was born, and his mother disappeared after being kidnapped by slave raiders. Carver was an infant at the time and was also kidnapped but was returned to the plantation alone. He was frail during most of his childhood and, as a consequence, was spared hard labor. Instead, he studied plants, grew medicinal herbs, and learned to love the earth and its harvests. Schooling was what he wanted, but education was not a reasonable option for a black child. Carver learned to read and write at home and then attended Simpson College.

Although his intent was to major in the arts, he was invited to transfer to Iowa State College of Agriculture and Mechanic Arts, where he began his career as a horticulturist. He became the first black faculty member at Iowa State and later taught at Tuskegee. At Tuskegee, he explored crop rotation systems that replenished the earth, which ultimately led to surplus stocks of peanuts. In response to the surplus, Carver developed hundreds of uses for the peanut, goldenrod, pecans, and corn stalks, to name a few. His love of the earth and his agricultural genius are a legacy that situates freedom and agency in the soil, which many African Americans despised because of their involuntary bondage in crop fields.

The Conversation

I was tired and needed a space to stretch and physically reorient myself. I sensed that the conversations were betwixt and between dream spaces and brain synapses, so my physical exhaustion was inexplicable. And yet, my body was achy and strained. The voice that answered my unspoken musings was direct. "Your grandmother knew better than to live in her head without giving her body a chance to ride."

I didn't recognize the voice or the face until it slowly came into view. It was Audre Lorde, poet, activist, and feminist. I read her poetry during college, including the story of her fourteen-year battle with breast cancer, so I couldn't help but wonder what direction the conversation would take. I could tell that she had a body, but I was hard-pressed to describe it. In the place of expected shape, I sensed strength and presence. The best way to describe Lorde as she appeared to me is "fully embodied." Everyone in the life after life seemed to have a discernible spiritual contour. But, because I did not carry the same memorized expectations for her that I that I had for the others (I had only seen her on book cover jackets), I could not bring her into focus.

What I perceived was a body that knew how to laugh and lament. There was none of the "normal" disassociation between head and heart. I wanted to get the conversation started with a discussion of her health issues and the implications for liberated bodies. To be truthful, I wondered why anyone battling cancer would take on the medical establishment as she had done. Just when she needed them most, she decided to question long-standing presumptions about their infallibility and healing power.

While I understood that Lorde was not the type of woman who would let the inaction that sometimes accompanies illness affect her prophetic discourse, I did wonder whether fighting medical systems while undergoing treatment was the best choice. I also wanted a better understanding of what it means to be "free" in the body when health is broken. I didn't know where to begin, but of course she heard that thought and responded.

Lorde I didn't have a choice. Some battles are chosen, and others come to you after the battle is already under way. When I became ill, it was as if I had entered a world where my embodiment as a woman could stand in the way of my healing. The doctors didn't know me or my spirit, which I considered to be a fatal flaw in our new bond to try to save my life. Even more vexing was the fact that they came to the healing task with preconceived notions about my personal bodily responses to their proposed cures. Most of the options that they offered to me were based on group statistics.

I understand that when medicine makes generalizations regarding race and gender, they do so hoping for the broadest possible application of new drugs. This may change because of recent biogenetic information about the uniqueness of individual, ethnic, and gender responses to drug therapies. In my instance, I wasn't trying to change the entire medical community or their ordinary procedures. I just needed for them to see me as a unique individual who was at the same time deeply embedded in various communities of care.

It didn't take a rocket scientist to determine that if my life was hanging in the balance, then the medical procedures that I received should include my resources for spiritual care and self-care. To maximize my prospects of survival, I had to make my community of support visible. I also had to live out of the same spheres of integrity and purpose that governed my days of health and wholeness. We all live in a web of inconvenient interconnections. I refer to them as inconvenient because relationships are difficult. Religious folk who pretend that they aren't don't get it.

What I realized was that the emotional distance that protected the medical providers might be detrimental to me and that my own sense of radical individualism might isolate me from the healing power of love that was being offered by family and friends. I stood rather bravely on the precipice of the unknown, knowing only one thing for certain: that my submission and silence would only hasten my demise.

I had cancer, but I also had hope. Imagine waking up and finding yourself dangling from a twenty-story building by your fingertips. You don't know how you got into the predicament, and you may not have the strength to pull yourself to safety, but you don't have to let pigeons nest in your hair. Over the long haul, I realized that I would not survive the cancer, but none of us survives life. I was hanging by my fingertips on the precipice of a great demarcation. I had done nothing to create the situation, and my prospects of being hauled to safety diminished with each "treatment," but given that reality, the illness was becoming a roadway toward a great adventure. It was becoming a canvas for mapping new life-giving possibilities for my future and for those who would journey after me.

As much as I loved life, I began to approach the end of my life as a comma, a deep breath at the end of an exhale. Something would come next, and the fact that I could not conjure the outcome or find it in the public library thrilled me, because the fact of its mystery assured me that I was facing a worthy and probably astounding outcome. I am purposely avoiding theological language because the discourse seems so small. If you cannot abide mystery, then you will put a period on the ends of open thoughts and you will not be able to entertain the potential for transcendence.

Why do we fear mystery? It is very freeing because of the questions that it invites. Anything that I can completely understand presents a future with specific limits. I wanted to be walked to the door of the unknown without a map drawn by folks who had never ventured past the doorsill.

I was still stuck on the image of pigeons nesting in her hair when a compact and precise gentleman entered the conversation. The best way that I can describe him is to note his presence as a source of localized power. One had the sense that if you plugged into his interiority, you would have more power than you needed but he would not be exploding all over the place. It was George Washington Carver.

Now, I like peanuts as well as anyone, but I never had much interest in Carver. There was (rightly or wrongly) my perception that he was a superannuated house Negro, the perfect servant who knew his place and made strides within those limited boundaries. My assessment was that he would never have led a revolution and might even have betrayed one for the "good of the people."

Carver Have you lost your mind? [He spoke with gentle annoyance.] I don't know you, lady, and it is clear that you don't know me. Just because I wanted to give working-class people dignity and skills does not make me a snitch, a betrayer of the brave revolutionaries. Unless you just insist on having your prophets robed in flowing white and bearded like Moses, I'd say that I fit the bill pretty darn well. People of color are dying from the contaminated food that they are eating, dying from the lack of meaningful work, displaced because of their disassociation from the land. Unless I miss my guess, I called these issues early on.

Not wanting to get into a verbal exchange, I sat silent but thought to myself, "Sorry, Sir," knowing that he would hear it. He nodded his acknowledgment and acceptance of my apology and turned toward Lorde.

Carver Sister, it's fine to engage the medical establishment on its gender bias and racism, but you were fighting on the wrong level. By the time that a person is seeing a doctor, something is already wrong. Doctors can only attempt to cure what has already been compromised. The real battle begins when we can assure people that the food that they eat is not contaminated with the chemicals that compromise health, when we can reintroduce them to the land as mother and provider, and when we can redefine "meaningful work" beyond Eurocentric capitalistic limits.

Lorde Well, if it isn't Carver. Of all the people to put me in dialogue with, you give me the peanut man.

Carver Watch out now, Lorde, I may be a peanut man, but you stole your famous quote from me.

Lorde How did you decide on one quote with all of the books I've written? I'm prolific, Brother, I did not have to steal from you. All of that hoeing in the sun must have affected your thinking. [They were joking and matching wits.]

Carver I don't have anything against hoeing or hard work. I was in the laboratory trying to turn the product that was hoed in the fields into a future for our people. But you're trying to get me off track [laughing]. Didn't you say, "You can't dismantle the Master's house with the Master's tools"? Wasn't that you?

Lorde Sure.

Carver Well, I said it first, if not in those exact words. I said don't depend on the ruling class for jobs, food, or esoteric education. Those are houses you can't live in. Why waste time dismantling when you can build your own with that same energy?

Lorde [amused] OK, peanut man, I don't mind sharing an antiracist thought with you, but the words were mine.

Washington So you admit that I'm the father of that idea.

Lorde I admit nothing. [They were both laughing.]

Carver Our discussion today is about the body, and as far as I am concerned, the peanut and the body are related.

Lorde OK, Carver, you did amazing work at Tuskegee, and at the time it was a bastion of agricultural innovation and health focus. The National Minority Health Foundation grew out of National Negro Health Week, which your colleague Booker T. started in 1915. I know my history, Brother. You even got the United States Public Health Service to support it.[4] So, I think it's pretty ironic that with all of your emphasis on health, peanuts make folks allergic these days.

They were enjoying themselves so much, I wondered if we would ever get back on topic.

Writer Let's keep the conversation going with a discussion of liberation in an individual context.

Lorde Fine by me. But let's keep it real. This generation needs clear advice. We crave freedom no matter what state we are in, but we tend to look for it outside of ourselves. We have the potential for liberty in every cell of our bodies, but we have to invite it to surface, claim it, and turn our intentions toward a different type of existence. Liberation is cosmic, but it is also very personal. Too often, we rely on freedom's reputation, the impact of its history, the songs and patriotic patter, and ultimately find that we don't have what we need. Instead, our lives are cluttered with a lovely bunch of ideas, and the memories of brave women and men who tried to wrest the concept from the ethereal to the pragmatic, but not much else.

Carver Well, it's really up to us to give the idea content. African Diasporan people are hungry for the fulfillment of freedom's promise. I believe that if the community would turn their hearts and minds toward the difficult work of carving out economic stability, they might find what they have always sought.

Lorde That's a fine thought, but the lived reality of a community often collides with dominant cultural ideas about who is worthy to possess such freedom. The politics of gender and sexuality continue to influence the effectiveness of the various freedoms that are available to us, and we tend to absorb and manifest those realities in our bodies. When I was struggling with cancer, I often found myself in great pain and despair. The question that I posed to myself was whether those feelings were an end product of the cancer or whether the disease released those hidden emotions.

If I were to attempt to identify the source of those feelings, I would say that as a gay black woman, I resisted and simultaneously absorbed the rejection and disapproval of organized religion and its commonly accepted critique of my body. When a society either implicitly or overtly allows the accumulation of toxic residuals of systematic social hatred, the end result will be deleterious. If it is a cosmological fact that we are all connected, then anger toward a group or an individual will eventually be lethal to the hated and the hater.

The religious authorities who teach love, forgiveness, and "don't let the sun set on your anger" are the same folks who denied my reality. So what does freedom mean for the varied and diverse communities who are in hot pursuit of its benefits? In the 1960s, we defined freedom as if all members of the black community held the same views and wanted the same things. We know better now. Freedom is complex; it is not a one-size-fits-all option. Instead, it tends to be more like access, and some of us have it and others of us don't. Phew, that was a long rant. I must have been holding that in for a long time.

Carver No argument there, Audre. But I wasn't addressing the diversity of those seeking to be liberated; instead, I was offering a resource that could be adapted to the needs and differences of the community. I know that my life and times included patriarchal perspectives, but the freedom that I was suggesting would have allowed egalitarian access to economic security through the reclamation of the land. It would still work today, especially for women in agrarian societies who predominate in the production of food. Real power is found not in temporary access to boardrooms but in control over the life-giving tasks of food provision.

Lorde While it may be true that the ability to feed yourself is freeing, it is also the case that the power to till, reproduce, and run families does not always translate into the power to save the lives of women and children. Women on the continent of Africa have been on the front lines of food production for centuries, but this treasure of knowledge did not help them when HIV/AIDS devastated the continent. Too often, women are offered liberation as a clay pot with many cracks and fissures, and those cracks, no matter how small, defeat the purpose of the vessel.

HIV/AIDS is killing a generation of people while prevention and cures are tangled in the politics of access. I don't know who benefits when an entire continent is allowed to die, when an entire generation of children will be orphaned, but I believe that this is a solvable mystery. Why are the treatments coming so slowly to the African continent? All we need do is follow the interests. Who has an interest, who will benefit, and who are the ultimate stakeholders?

Writer Does liberation have a set definition, or does it change to fit each crisis? And, are there points of agreement as to how we will reach our goals?

Lorde Good questions. I have noticed that those who participate in the struggle for justice seem to take such ownership of the process that other seekers have to joust for space. Each justice-seeking group in the mid-twentieth century wanted to believe that their liberation movements (focused on race and ethnicity) exemplified the pinnacle of all justice pursuits. Hence, the resentment toward the post–civil rights activism of feminists/womanists and the GLBT community.

This narrow approach to liberation as the private domain of one oppressed group over another is subsumed in the statement attributed to the Rev. Bernice King a few months before she participated in an antigay march in Atlanta. While speaking at a church in Auckland, New Zealand, in October 2004, she is reported to have said, "I know deep down in my sanctified soul that he [MLK] did not take a bullet for same-sex unions."[5]

By contrast, in a speech four days before the thirtieth anniversary of her husband's assassination, Mrs. Coretta Scott King said, "I still hear people say that I should not be talking about the rights of lesbian and gay people and I should stick to the issue of racial justice. But I hasten to remind them that Martin Luther King Jr. said, 'Injustice anywhere is a threat to justice everywhere.' I appeal to everyone who believes in Martin Luther King Jr.'s dream to make room at the table of brother and sisterhood for lesbian and gay people."[6]

Coretta's statement did not presume that the liberation that Martin Luther King fought and died for was a one-strand, one-note quest for one people. She instinctively knew that even if you want to keep liberation to yourself, you can't because it will sneak over and hug your neighbor and enlighten the teens on the local basketball court.

Carver It is clear that liberation is an event that has many dimensions. We can't forget the liberation of human bodies through good health. You mentioned the HIV/AIDS pandemic, and I am wondering how that plays out in liberation discourse. While I have great concern about the consequences of a disease that devastates a continent, I am not certain that the liberation project is logically connected.

Lorde Nothing could be more connected, Carver. Freedom is difficult to inhabit when you are sick. Health is spiritual as well as physical, and sometimes our sense of powerlessness is the result of a failure to draw from deep wells of shared wisdom. We can wait for experts to solve all of the problems of wellness in the community, or we can use the resources of the medical establishment, the healing wisdom of the elders, and the experiences of neighbors and pool resources. We have always known how to care for one another. The HIV/AIDS pandemic requires distribution of the latest drugs, but it also requires re-education, discussions of sexual politics, discussions of theodicy, and a willingness to offer comfort and care. To do otherwise is to present our bodies as living sacrifices in the worst possible meaning of that scripture. I want to redefine freedom as health and wholeness.

Carver All right, maybe I'm catching your drift. If we accept other people's stories about who we are, we will have nowhere to anchor our hope or our health. We will be adrift.

Lorde That's right, Carver.

Carver The task is to be whole in body and soul in the midst of a fragmented society, "healthy" in a country that eats itself into illness. I want to urge oppressed people to begin to think of their bodies as gifts that must be honored. We would not receive a plant as a gift and then purposely let it die by feeding it all of the wrong things and neglecting basic requirements for survival.

Freedom is for every member of the earth community. A processive view of freedom does not focus on human interests to the exclusion of the rest of earth's inhabitants. No longer can one sing, "Oh, Freedom, oh Freedom, oh Freedom over me, and before I'd be a slave, I'd be buried in my grave and go home to my God and be free," and then immediately attend the church dinner to consume chickens enslaved all of their lives, debeaked, feed dead ground kin along with antibiotics, growth hormones, and who knows what else. Listen to what Peter Singer says.

> We share with [animals] a capacity to suffer, and this means that they, like us, have interests. If we ignore or discount their interests, simply on the grounds that they are not members of our species, the logic of our position is similar to that of the most blatant racists or sexists who think that those who belong to their race or sex have superior moral status.[7]

Lorde I really hadn't thought about that, but thanks for making lunch problematic.

At this point, another woman entered the conversation. She was hale and hearty with a presence that made you pay attention.

Hamer Wait a minute, Carver, I never ate the synthetic foods that you are talking about. What I cooked still had Mississippi dirt clinging to it.

Lorde If it isn't Fannie Lou Hamer! Girl, I thought that you'd be in the chapter with Martin and the others.

Hamer That was the original plan, but I realized that I'd never been given the opportunity to talk about my health or the difference that a broken body makes in the abbreviated stories of our people. I will stop by later to say a word to Martin. But this chapter is where the rubber meets the road. People remember me saying that I was sick and tired of being sick and tired, but they don't think of it as more than an adage or saying. Actually, I was sick a lot of the time. I had hypertension, diabetes, and cancer, and my body was subjected to every manner of abuse, from beatings to unwanted operations.[8] The things that I have been through would shock you, peanut man . . . and I mean no disrespect.

Carver None taken.

Lorde Fannie Lou, we are talking about the lack of spiritual and physical connection to the earth and the universe and the need to reconnect to one another as a way of redeeming the health of communities. Now that I'm really thinking about it, the difference between discussions of freedom during the movement and discussions now shift dramatically toward individual attainment.

Hamer That's right. King talks about a societal healing, education for all, and the end of conflict that originates out of discomfort with difference. Today, we talk about the right to be full participants in a consumer culture. We are encouraged to be loners, but my advice to this generation is to remember that they are not alone.

> The group is within each one of us. . . . I sense my family, the body of my nation, and sometimes the entire earth in my own body. When my nation suffers, I suffer. We are not independent cogs. . . . The well being of the group is our own well being.[9]

Carver Issues of survival have always impeded the difficult work of shaping a communal identity. My ideas have been misrepresented so often, but all I was trying to convey was that if you are deemed to be a servant class, then be the best on the face of the earth and learn the skills that will support the survival of your own people.

The land didn't enslave us, the cotton didn't force us to pick it, and the peanuts didn't demean us. Why reject the land that gives us real freedom? To depend on corporate interests for food when the earth will supply it seems to be another form of voluntary bondage. What will you do when the trucks stop rolling into your local supermarket, when the shortages because of the abuses of the land begin to occur? Will you remember that a few seeds and some water will keep you free from dependence for sustenance?

Hamer Becoming the best possible servants is fine if it's a choice. But when an entire racial/ethnic group is forced into servitude or when inferiority becomes their defining quality, then there is no glory in perfecting a response to that abuse.

Carver I see your point, Fannie, but you are missing one of mine. If we are talking about freedom, I would argue that there is freedom in farming.

Lorde We understand your point, Carver, but not even the seeds are free anymore. Monsanto and other companies have patents that will not allow farmers to collect and save their seeds.[10]

Hamer But even if this basic right was reinstated, liberation that emerges from the land requires more than a few seeds and some water. It is backbreaking work. I think that it will be several generations before the land becomes a site of communal return, because the trauma of forced labor caused so much physical and psychic pain. If I make you hoe every day, when you escape you are going to avoid the site of your enslavement for a long time. The land was the only option back in the day, and some of us were supposed to be inventing, or writing, or negotiating, but we were all picking peas or something. Give it a chance, George. Some will return.

Lorde I want to talk about the politics of a rural communal focus that you seem to advocate. There was, and still is, a mind-set in some rural communities that rejects difference. It probably has less to do with the people than it does with the limited opportunities to host difference. It's the cities that receive new immigrants and ethnic shifts. In rural settings, people sometimes live and die in the same place without the opportunity to journey geographically or to meet other members of the human family. Without the friction of new languages, faces, and food choices, people construct closed comfort zones.

Hamer I'm not certain that you are right about that. I mean, that's how it seems from the outside. But I know in our rural town we had enough difference to make your head spin. Now we didn't always handle it well, but it was considered to be just a natural part of life. In fact, the conformity of thinking and behavior that prevails today just didn't exist. Everybody was quirky in one way or another, and certainly there were enough lifestyles to choose from, but we didn't characterize people in that way. You might be in a field alongside women who could whip any man on the cotton row and men who could sew you an outfit that would have you strutting down the aisle on Easter Sunday morning. We accepted members of the community no matter how odd we considered their choices or behavior to be because we relied on solidarity to sustain us in the midst of a battle for survival.

Lorde But how did you handle the issue of sexual identity?

Hamer Not very well. As a people, we were invisible to the dominant culture based on color differences. When we encountered difference of lifestyle within our own community, we tended to compound an already opaque existence by offering a second veil of invisibility to shield all of us from conversation about difference. If pushed, we would have blamed our lack of unconditional acceptance on the Bible. Talk about layering one lie on another. Peter J. Gomes says it well.

> I happen to believe that our attitudes toward homosexuals and homosexuality are not based on the Bible, although they appear to be because of the way in which we read the Bible. Those views

come from the culture. Cultures however, come and go: they are not absolute nor are they infallible. . . . These matters need the most careful and thoughtful scrutiny in light of Christian black America's conspicuous "biblical homophobia." Our perceived self interest—that is, the preservation of our own status by keeping at risk the status of another—is contrary to our own best interest, which is to become not simply liberated but liberators.[11]

Lorde The weird thing is that our treatment of sexual preference has no historical relevance to our African origins. We have adopted a Eurocentric aversion to the body that has no relevance to the wide range of body identifications on the African continent and in ethnic and Diasporan communities everywhere.

Writer This is a wonderful, freewheeling conversation, but can I suggest that we cover just a few last topics like the politics of food and the science connection?

Hamer Sure. To my way of thinking, social choices are influencing the politics of food and health in the community. If you believe that you must work three jobs to own the latest technology, or to keep the credit card bills paid at minimum level, then you will eat what you can when you can.

Carver And there won't be any Mississippi dirt on any of it. How can we believe that our bodies will thrive on tomatoes that never rot, carrots that stay fresh for months, genetically engineered foods, and artificially enhanced milk. From what I understand, the twenty-first century has some of the fattest people in the world who are starving to death for lack of nutrients and vitamins.

Lorde But don't you think that fast-food eating habits are related to a shift in the identity of all Western folks? If capitalism has convinced all of you that you are worker/consumers, and the vast majority of your time on earth is spent toiling in one corporate field or another, then it doesn't matter what you feed a cog in the marketplace wheel. To be fully human, you must take time to play with your own children, to grill the bass you just caught, to sit on the porch and contemplate the meaning of life.

Carver If you don't have time to eat, you don't have time to live. If you are decentered physiologically, it's only a matter of time before everything is out of kilter.

Hamer It's no wonder that "terror" is on everyone's mind. A skittish, fast food–starved society will be very susceptible to fears real and imagined. I have not talked about it very often, but my body was the site of a terrorist attack. That's what I call it anyway. It was 1961, and I was picking cotton and trying to eke out a living in the Delta. I went to the hospital to have a "knot" removed and came out completely stripped of my ability to have children. They gave me a hysterectomy for no reason. I would never have known if rumors had not circulated in the "big house" that eventually got back to me.[12] When I confronted the doctor, he said nothing, but others have said plenty about the reproductive rights of black women.

> We don't allow dogs to breed. We spay them. We neuter them. We try to keep them from having unwanted puppies, and yet these women are literally having litters of children.[13]

This statement and the action of the doctor who took away my options for motherhood are so insanely evil that there are no words to describe how I felt.

Carver I don't mean to offend, but are you certain that there was no medical reason for such extreme measures?

Hamer Man, because we're both in a better place, I'm not going to jump on you. Let's just say that you're right; don't I have the right to make the decision?

Lorde You are outnumbered peanut man, back off. Girl, I am so sorry. I know that sorry doesn't mend much, but I want you to know that I care.

Hamer In a way, you get what you need in life to shape and prepare you for what is to come. That thing broke me up, but it didn't break me down. Years later, when they were beating me in Mississippi, I could take it because my body had already been broken from the inside out. I watched them wailing away until their arms got tired; then I talked to them about God. It scared them half to death. No tears, no pleas from me, just prayers.

Carver The only way that those types of atrocities can occur is if those committing the offenses are alienated from themselves and the earth. One can live without connecting to the cosmos, but the price is an emptiness that will allow you to harm yourself and your neighbors. I hate to keep harping on the same thing, but you can't grow a peanut in the time that you go through a fast-food drive-through window. Nurturing the food that will nurture you takes time, and while you tend and wait, you have time to live.

Hamer Not many people are paying attention to the politics of food. We watched the takeover of the family farms a few years ago, the growth of corporate food factories, and the pollution of waterways and the earth. Then we shifted from eating locally to depending on centralized food sources trucked in from all over the nation. This was happening while the social protest movements were trying to solidify a more equitable social order.

Lorde The task is to keep your eye on the things that will affect the health and welfare of the community. We might get integrated schools, but if the lunchrooms serve lunches that dull rather than ignite potential, then what have we accomplished?

Hamer I'm not so certain that the neglect of the body, the corruption of the food supply, and the advances in science that are seldom taught to the poor are unrelated.

Writer Well, this may be the time to take the turn toward science. In this chapter, we are dealing with biogenetic changes, liberation, and . . . posthumanism.

The silence was sudden and profound, and then they all said at the same time "posthumanism"?

Writer Yes. Posthumanism suggests that the bodies that we have now may be transcended by neurological and technological interfaces with computers. It also concerns germline therapies and cellular alterations. We are talking about inheritable genetic changes in a person's reproductive system that affect not just the individual but also future generations.

According to Robert Pollack, "the DNA that succeeds in traveling from the egg and sperm of one generation to the egg and sperm of the next is called the germline. We may say with some confidence that the lifetime of a species is the lifetime of its germline."[14] Some scientists and theorists suggest that modifications of the human germline will have profoundly permanent consequences, including the possibility of a posthuman or transhuman species.

Determining the exact meaning of *transhumanism* or *posthumanism* depends to a great extent on who you ask. For scientific advocates of posthumanism, the terms presume the ability to transform the human species through germline modification or other technological or genetic enhancement.[15] N. Katherine Hayles makes no such presumptions. She offers a phenomenological view of posthumanism.

> In the posthuman [world] there are no essential differences or absolute demarcations between bodily existence and computer simulation, cybernetic mechanism and biological organism. . . . The posthuman subject is an amalgam, a collection of heterogeneous components, a material-informational entity whose boundaries undergo continuous construction and reconstruction.[16]

Lorde Well, this is certainly not something that I've thought a lot about, but the positive potential of posthumanism may be eclipsed by the human penchant for evil. Given our past history as a species, one can reasonably wonder whether the trans/post/human state will provide fertile soil for abuse.

Writer I'm talking too much, but one last definition is necessary. Although I am using the terms interchangeably, there are those who see transhumanism as an intermediary step between the human and posthuman state.
According to the Center for Genetics and Society, the study of transhumanism began in the 1990s as a biogenetic libertarian movement.

> Transhumanists view human nature as a work-in-progress, a half-baked beginning that we can learn to remold in desirable ways. Current humanity need not be the endpoint of evolution. Transhumanists hope that by responsible use of science, technology and other rational means we shall eventually manage to become posthuman, beings with vastly greater capacities than present human beings have.[17]

Although the pathway toward this biotechnological future is strewn with promising scientific breakthroughs and science fiction conjectures, debates about limits and goals are highly polarized. Some point with despair to humans "playing God"; others warn that the "essence" of humanity cannot be captured in a laboratory. To those who seek the ultimate fulfillment of human potential in the science of germline modification, the future seems to be a limitless frontier.

Carver But what can we really say about the future? Certainly, we can extrapolate ultimate outcomes based on advancements upon current knowledge, but we can no more predict the future based on current knowledge than Edgar Allan Poe could have foreseen Microsoft Word. Despite all of our projections and hard work, the future can still surprise us, and I expect that it will.

But it is not only the future that can surprise us. This whole subject will be a surprise to the average person. Most will think that you are dealing with science fiction.

Hamer If trans/posthumanism is our future, it will inevitably have the most impact on those communities and stakeholders that know the least about its promise and perils.

Writer Yes, bioconservatives, such as Francis Fukuyama or Leon Kass, are concerned about the undermining of human dignity and the harm to essential aspects of human life as well as the potential for violence based on the perceived superiority of one human group or another.[18]

Lorde Well, it's clear to me that if that's all they are afraid of, then they don't know how dehumanization works in our society. Don't take my word for it. Members of communities marginalized by race, sexuality, or gender will tell you: they are already suffering from each of the enumerated "future dangers." Moreover, they would be perplexed to find that the issues of oppression that they struggle with daily are being projected as potential dangers of a posthuman future. The marginalized are not only left out of the dialogue and visioning process; they are suffering the so-called "future harm" now.

Hamer This is my problem with all of this futuristic stuff. We don't have to wait for posthumanism for stories of abuse. I was not the only one who endured a forced sterilization dubbed a "Mississippi appendectomy." Devastating poverty affects the growth and learning capacity of children, which in turn affects their ability to successfully raise their own children and escape the cycle of poverty. The effects of malnutrition and the lack of access to health care are as catastrophic as the potential genetic alteration of human cells.

The instances of residual social effects on the human germline of these marginalized populations are too numerous to count; however, even more disturbing are the intentional uses of state and social power to affect human potential at the most foundational biogenetic levels. George, your alma mater knows more about this than most.

Carver Will we ever live down the Tuskegee experiment? Although it was a heinous example of medical racism and self-loathing, it was not the only experiment of its kind. There were many other purportedly scientific experiments that subjected African Americans to medical procedures that could affect the germline.

> For decades the recommended dosages of x-rays administered to African Americans were usually one and a half times, often double, those administered to whites. . . . The informed consent from individual African Americans would entirely miss the point about the consequences variably affecting the fate of this group,

as members of a group. . . . The effects of systematically higher doses of x-rays on African Americans may actually concentrate germ-line effects on that group.[19]

As technology advances, the questions proliferate. Given past history, can the science community be trusted to handle such critical issues in ethical ways? Should we risk the opening of a genetic Pandora's box for potential cures that lie just beyond the current limits of scientific knowledge and ability?[20]

Lorde But how can we trust science and medicine when history tells us that this is not prudent? We want scientific advances that liberate. "A technology is not justified if it neglects the poor or vulnerable, or if it widens the gap between the haves and the have-not's."[21]

Carver I think that I can solve this. Since we know that race is a social construction and does not exist as a legitimate biogenetic category, why not allow this idea of posthumanism to help us transcend race by biogenetically engineering racial neutrality into posthuman beings? They would be devoid of the physical traits deemed to be inferior.

Hamer He's kidding, right?

Lorde If you are trying to get rid of race through biogenetics, forget it. Scientists offer prescient warnings that the idea of race cannot be so easily discarded. Genetic research indicates that while we share 99.9 percent of the human genome, and theologically we believe that we are part of the same spiritual family, race may still affect the health and future of the two-thirds world.[22]

Another wave of the future is the matching of genetics with specifically tailored medications, or pharmacogenetics, for racial/ethnic groups. This tailored approach to health care makes everyone nervous because of a racist past. But if we leave race out of the genetics and pharmacological research, we may be excluding people of color from the benefits of recent scientific breakthroughs.[23]

Hamer This is not the only potential harm. One wonders whether profit-motivated companies will spend the same amount on research for the Zuni as they do for white males who control Congress and usually run the biotech companies.[24] Maybe they will, and maybe they won't. The fact that we are unsure places ethnic people at risk, whether by abuse or neglect. Indeed, to partake of the future benefits of science requires that people of color trust the scientists, something they have learned not to do.

Carver Although it may seem far-fetched, the debate as to whether stem cell research or germline manipulations violate potential personhood or sacrifice potential humans for already living humans evokes the utilitarian rationale of the Tuskegee experiment. It could be argued that from the perspective of the progeny of slaves, control over the human germline differs only in scale from the control that slaveholders wielded over the bodies and the procreative destinies of their captives.

Lorde We have learned from critical studies of race and from the analysis of "whiteness" in North America that white privilege is an unacknowledged and taken-for-granted benefit and an invisible but immutable legacy. This legacy gives dominant cultures power over dark, female, physically challenged, gay, and lesbian bodies. The question is whether we (and by "we" I mean the human community) want to vest the power to genetically alter the human family in systems that have been steeped for decades in the malignancies of racism.

Hamer I vote no. If we really want liberation, we will have to realize that technologies are seldom neutral in application or intent, when they are being used by socially located human beings who have inculcated conscious and unconscious cultural biases. As Philip Bereano notes, "We are not all in this together." In Western societies, a small group of people, usually white and male, makes the decisions and controls the financial resources that determine the scope and application of new technologies.

Writer I know that it's a lot to cover, but we haven't talked about the violence factor. Posthumanists acknowledge the potential for violence between original and altered humans. Nick Bostrum warns:

> The new species, or "posthuman" will likely view the old "normal" humans as inferior, even savages, and fit for slavery or slaughter. The "normals" on the other hand, may see the posthumans as a threat, and if they can, may engage in a preemptive strike by killing the posthumans before they themselves are killed or enslaved by them. It is ultimately this predictable potential for genocide that makes species-altering experiments potential weapons of mass destruction and makes the unaccountable genetic engineer a potential bioterrorist.[25]

Lorde Germline modifiers as bioterrorists! This is strong language with prophetic overtones. In this present age, it is unfortunate but true that the ability to wage successful wars helps to determine where power lies, who has worth, and who is at the bottom of the value pyramid. A species split between humans and posthumans would shift tensions away from today's rigid polarities of race, gender, and class. Although this would be a long-awaited respite for those who have been under siege, it would also be a continuation of the politics of difference and fear.

Carver The other issue in this discussion of germline modification and liberation is whether anyone can be free if parents can manipulate the lives of their children. What does freedom mean if another human being determines gender, height, and eye color?

Hamer Can we reflect the diversity in the Creator's original plan if those deemed inferior, insane, unfit, and weird are engineered out of existence? Historically, racial/ethnic children have languished in foster care while white babies were sought from every corner of the earth. It is not unusual in this century to see families with children adopted from nations around the globe. Given the right technology, it might be possible to engineer problematic cultures or surplus human beings out of existence. I realize that I am describing an unlikely application of genetics, but who would have imagined posthumanism?

Lorde Unfortunately, almost everything is within the realm of possibility, and that worries me. What will happen to the gift of difference embodied in idiosyncratic, physically or mentally challenged bodies in our midst? In non-Western cultures, those who exhibit oddities and traits that don't comport with societal expectations have been given places of honor. In fact, some ethnic people rely on the "odd members" of society to be the shamans, priests, and mediators of the divine. One must wonder how a John the Baptist, known for wearing skins and eating insects, not to mention his ranting sermons, would fare in our Western culture. Or Isaiah, lying naked, or Jeremiah, weeping and publicly imploring God. If freedom has any meaning, it must allow us to deviate from the "norm."

Writer My own brother has a serious mental illness. I know that his engagement with the world differs significantly from mine, but it is as crucial to our community as any other. Those who barely hang on, who like the lilies of the field (to paraphrase Jesus), take no thought as to what they should eat or drink, savoring and suffering their days. Should they be engineered out of existence?

Carver We are not in any immediate danger of this occurring, but we are in danger of losing the meaning of God's liberation. Freedom in its deepest sense is not political, social, genetic, or biological. Freedom comes from the affirmation that we are connected to something greater than ourselves. As we look inward, we must also look beyond merely technological solutions.

We are struggling not against flesh and blood but against and in the midst of geopolitical, scientific, and spiritual forces that will put our communal resolve to the test. Germline modifications will not save us from our own evil acts.

Lorde Biogenetic creativity should be driven or shaped not by privilege and power but by the universal longing for freedom, most intensely present among those who have been excluded from its fulfillment. This longing must be the living matrix and reference point for applications of new human enhancement technologies. And finally, scientific objectives and core beliefs about posthumanism must take into account the perspectives of the excluded, even at the cost of a far more complex account of our technocultural situation.

Hamer Perhaps the good news for all people must include the stories of genetic possibility.

Writer Perhaps it is in the gaps of our theological and scientific knowledge that the fullness of humanity can be found. Genes constitute only a small percent of DNA. Gene-coding regions are interspersed among billions of non-coding DNA bases whose functions are still largely unknown. We have presumed in the past that these unknown areas have no function. This is an arrogant dismissal of human/divine space. I embrace these vast areas as the blessed unknown, a place where spirit can dwell, liberty can be nurtured, and our multiple depictions of God can be entertained.

Lorde There is another way of understanding posthumanism. With our wars, petty assaults, and social dysfunctions, we cannot at this point claim to be fully human. Yet, each faith tradition has been given a human/posthuman exemplar in Christ, Buddha, the Prophet Muhammad, and others. To "follow" or live out the mandates of peace, love, and communal responsibility could be considered posthuman for all who lament our current failings. This is a processive opportunity for grace and renewal, for "the story of every living thing is still in the telling."[26]

Hamer It comes down to the definition of freedom. Are we are entitled to do all that is within our reach, and are we free to extend that reach by any means necessary? In a sprint toward a future that remains shrouded from our view, we have not considered long-range consequences, nor have we included the marginalized in the discussion.

Carver There is only one human story, one genome with many interwoven narratives, but we cannot construct a future without the inclusion of counter narratives. Experiential stories of suffering offer an alternative history of past "inhuman" experiments that maimed and killed. In every generation, we attempt to vault beyond currently available knowledge to that realm where what is possible merges with probability, scientific creativity, and social context.

Lorde Scientific advances must also take into account a cosmos that is indeterminate and inclusive, mysterious and multidimensional. An expansion of enlightenment epistemologies dressed in the new guise of posthumanism will not serve us well.

As Leon Kass notes:

> Human nature itself lies on the operating table, ready for alteration, for eugenic and neuropsychic "enhancement," for wholesale redesign. In leading laboratories, academic and industrial, new creators are confidently amassing their powers and quietly honing their skills, while on the street their evangelists are zealously prophesying a posthuman future. For anyone who cares about preserving our humanity, the time has come to pay attention.[27]

Hamer If the "new creators" of posthumanism are on the streets prophesying, as Kass contends, they certainly aren't hanging out in my old neighborhood. My advice to this generation is to pay attention and inform yourselves.

Carver I am encouraging this generation to lay the foundation for a grounded liberation. To be cosmologically oriented does not mean that they must have their heads in the heavens. The universe is closer than your next breath.

Lorde It is in the soil of the earth and the soil of our souls. We are the cosmos and the cosmos is us.

Hamer Now ain't that good news!

Summary: How Is Freedom Embodied?

Writer The conversation is over, but it seems as if there is still more to say. What they seemed to be saying is that the body is ground central for liberation. This is not an intellectual exercise, but a reorientation of mind, body, and spirit.

Sarah Mind, body, and spirit are not separate elements of human existence; they are interlocking and overlapped. The wholeness of the universe is reflected in a holistic and integrated human reality.

Writer But how do we embody liberation with integrity? We have tried to wear, buy, and vote our liberation. Nothing has gotten close to the personification of freedom that reflects the reality of the cosmos.

Sarah The more you learn about the origins of life in the universe and its future, the better understanding you will have of the present. "The universe desires to show itself! The universe is a showing of the unnameable mystery out of which being shines forth."[28]

Writer So, we are here with specific destinies, but part of that destiny is the reflection of the mysteries of creation and the intelligibility of the cosmos.

Sarah I like the direction of the conversation. The discussion of posthumanism has to be cast in a larger context. I want to end with cosmologist Brian Swimme's discussion of the dreams of the universe as embodied in the human community.

> We are creating with our imaginations a period of rebuilding, where the intercommunion of all species will guide our life activities. We must come to understand that these dreams of ours do not originate in our brains alone. We are the space where the Earth dreams. We are the imagination of the Earth, that precious realm where visions and organizing hopes can be spoken with a discriminating awareness not otherwise present in the Earth system. We are the mind and heart of Earth only in so far as we enable Earth to organize its activities through self-reflexive awareness. That is our larger destiny: to allow the Earth to organize itself in a new way, in a manner impossible through all the billions of years preceding humanity. Who knows what rich possibilities await a planet—its heart and mind—that have achieved this vastly more rich and complex mode of life.[29]

Writer The universe dreams through me? If I can believe that my life is an expression of the imagination of the earth, then I have hope no matter the circumstances. Fannie Lou really had it right. This *is* good news!

The Exercises

1. Write an e-mail from your body to you. In the message, your body should tell you what it needs from you and the areas of neglect and affirmation. Answer the e-mail, and tell your body how you intend to

respond. (This exercise is adapted from Natalie Goldberg's *Writing Down the Bones*).

2. Write a two-page story about an encounter you have had with the medical community either through illness or healing. What surprised you most about the encounter?

3. Choose an elder from this chapter and summarize his or her approach to issues of faith and health.

4. What questions do you have about posthumanism? Give a brief narrative summary of three issues. What would you engineer in or out of the human species?

5. What has the universe dreamed through you?

For Further Reflection

George Washington Carver

Albus, Harry James. *The Peanut Man: The Life of George Washington Carver in Story Form*. Grand Rapids, Mich.: Eerdmans, 1948.

Burchard, Peter Duncan. *Carver: A Great Soul*. Fairfax, Calif.: Wise as Serpents Harmless as Doves, 1998.

———. *George Washington Carver: For His Time and Ours*. "Special History Study: Natural History Related to George Washington Carver National Monument, Diamond, Missouri." George Washington Carver National Monument, National Park Service, United States Department of the Interior. 2005.

Coil, Suzanne M. *George Washington Carver*. New York: F. Watts, 1990.

Elliott, Lawrence. *George Washington Carver: The Man Who Overcame*. Englewood Cliffs, N.J.: Prentice-Hall, 1966.

Kremer, Gary R., ed. *George Washington Carver in His Own Words*. Columbia: University of Missouri Press, 1987.

McMurry, Linda O. *George Washington Carver, Scientist and Symbol*. New York: Oxford University Press, 1981.

Merritt, Raleigh H. *From Captivity to Fame or the Life of George Washington Carver*. Boston: Meador, 1929.

Audre Lorde

Lorde, Audre. *A Burst of Light*. Ithaca, N.Y.: Firebrand, 1988.

———. *The Cancer Journals*. 2nd ed. San Francisco: Spinsters Ink, 1988.

———. "Manchild: A Black Lesbian Feminist Response." *Conditions Four* 2, no. 1 (1979). Reprinted in *Sister Outsider: Essays and Speeches*. Freedom, Calif.: Crossing, 1984.

———. "Notes from a Trip to Russia." In *Sister Outsider: Essays and Speeches*. Freedom, Calif.: Crossing, 1984.

———. *Zami: A New Spelling of My Name*. Watertown, Mass.: Persephone, 1982.

Lorde, Audre, and James Baldwin. "A Conversation between James Baldwin and Audre Lorde." *Essence*, August 1982, 58–60.

Lorde, Audre, and Adrienne Rich. "An Interview with Audre Lorde." *Signs: A Journal of Women in Culture and Society* 6, no. 3 (1981): 713–36.

Fannie Lou Hamer

Interview by Robert Wright, August 9, 1968. Transcript (41 pp.). Howard University: Ralph J. Bunche Oral History Collection Archive number: DC28.184. Interview by Anne and Howard Romaine, November 22, 1966. Transcript (29 pp.). Mississippi Department of Archives and History: Anne Romaine's Mississippi Freedom Democratic Party Oral History Archive number: OH 81-08.5.

Interview by Neil McMillen, April 14, 1972, and January 25, 1973. Transcript (47 pp.). In Martin Luther King Jr. Center for Nonviolent Change's Anne Romaine Papers, 1963-1969 Stanford University: Project South Oral History Program Archive number: 0491-1. Cassette. Transcript (17 pp.). Also in Southern Regional Council's Will the Circle Be Unbroken Collection University of Southern Mississippi: Civil Rights Movement in Mississippi Vol. 31.

(Bibliographic excerpts from the University of Southern Mississippi)

Mills, Kay. *This Little Light of Mine: The Life of Fannie Lou Hamer*. Lexington: University Press of Kentucky, 2007.

We know the road to freedom has always been stalked by death.
  —Angela Davis

7. Killing Me, Killing You, Killing Us!
Violence and Liberation

The object of one's hatred is never, alas, conveniently outside but is seated in one's lap, stirring in one's bowels and dictating the beat of one's heart. And if one does not know this, one risks becoming an imitation—and therefore, a continuation—of principles one imagines oneself to despise.
  —James Baldwin

Love is the only true revelatory power because it escapes from, and strictly limits, the spirit of revenge and recrimination that still characterizes the revelation in our own world.
  —René Girard

This chapter speaks to the issue of state-sanctioned and random violence and the possibility of communal healing. The conversation partners are Ida B. Wells-Barnett, whose life focused on resistance to lynching; Stanley Tookie Williams, a gang leader sent to death row who was executed by the State of California after his rehabilitation was well under way; and Huey P. Newton, whose Black Panther Party confronted the violence of the nation with communal resistance and empowerment.

The Elders

Ida B. Wells-Barnett, 1862–1931

Why is mob murder permitted by a Christian nation? What is the cause of this awful slaughter?[1]

Ida B. Wells-Barnett was known as a freedom fighter during an era when violence was the expected response to resistance. She was born in Holly Springs, Mississippi. While she was still a teenager, her parents died of yellow fever and she became the sole support of her siblings. She taught school until she began to write for a church-based newsletter in Memphis, the *Free Speech and Headlight*. Wells-Barnett also began an antilynching campaign after three of her friends (Thomas Moss, Calvin McDowell, and Henry Stewart) were lynched. They were grocers whose business was growing to the extent that it was affecting the financial monopoly of "white" grocers in town. When Moss, McDowell, and Stewart were attacked by angry white businessmen, they defended themselves, shot an attacker, and found themselves jailed. A mob removed them from the jail and lynched them.

Prior to this act of violence, Wells-Barnett had already assumed the role of an investigative journalist. In her column, she exposed injustice without any concern about her own safety. After the lynching, she wrote an article encouraging members of the black community to leave Memphis because of the level of unchecked violence. She took her own advice and resettled in Chicago. She is also known for refusing to be removed from a segregated trolley car. She fought as she was dragged off and then sued the railroad. She won on the lower level, but the judgment was overturned by the State Supreme Court. Wells-Barnett worked with W. E. B. DuBois to form the NAACP, and she opposed Booker T. Washington's accommodationist stance. She was a woman familiar with violence and its power to undermine personal and social stability.

Stanley Tookie Williams, 1953–2005

In my dreams I've envisioned my liberation many a times, as a matter of fact, I was telling an individual the other day that in my dreams, whenever I run into some albatross or some type of dilemma, I seem to float away from it and in my mind, that is a sense of freedom. That is a sense of avoiding, eschewing or shunning any type of madness.[2]

Stanley Tookie Williams was born in New Orleans, Louisiana. He is known as a cofounder, with Raymond Williams, of the violent Crips gang. According to Williams, they founded the gang as a superpower in the neighborhoods that would trump the small turf wars between gangs. The result was the exact opposite. The Crips became a crime syndicate that spread throughout the country. Williams was jailed for four murders that occurred during two robberies. Although Williams denied participation in the crimes charged to him, he refused to "turn state's evidence" against the Crips. On death row, although he argued eloquently for clemency, he was executed in December 2005. The execution by lethal injection was botched, and it took twenty minutes to administer the lethal injection.

By the time he was executed, Williams had become a public figure and a self-proclaimed reformed gangster who wrote from jail to convince young people to avoid his fate. His nomination for a Nobel Peace Prize and Prize for Literature for his children's books was generally discounted by the media because of the egalitarian process of those nominations. Williams was a complex individual.

Huey P. Newton, Ph.D., 1942–1989

As long as the people live by the ideas of freedom and dignity, there will be no prison that can hold our movement down.[3]

Huey P. Newton was born in Monroe, Louisiana, but grew up in Oakland, California. According to most sources, he attended school but was self-taught. This description refers to his persistence in improving his reading skills by immersing himself in the classics. During his attendance at Oakland City College, he met Bobby Seale. Together, they created the Black Panther Party for Self-Defense in 1966. The party represented the translation of their political interests and frustration with the abuse of African Americans at the hands of the police. Newton is often depicted with a rifle in his hand. He was studying law and knew that the open display of weapons was legal in his county, whereas concealed weapons were not. When party members began to patrol the neighborhood with weapons displayed, a clash with police became inevitable. In 1967, after a shootout, a police officer died, another was injured, and Newton was also shot. Newton jumped bail and sought refuge in Cuba, but he later returned to face the charges and was acquitted.

Newton's strategy to heal a community under siege was to provide support for the elderly and breakfast and education for the children, using an adapted Maoist model of social structures. The FBI went to extraordinary lengths to infiltrate and destroy the basis of the Black Panther community and national support. Those efforts are documented in public records and in Newton's doctoral dissertation. Newton earned a Ph.D. from the University of California (Santa Cruz) in 1980. Nine years later he was killed by a drug dealer under circumstances that are not clear. He died on the streets of the neighborhood that he had tried to protect.

The Conversation

It may seem that the stories have been easy to hear and to write. Nothing could be further from the truth. In this chapter, my heart was weighed down with anxiety and grief. There have been too many killings on the pathway toward liberation, too many family killings of the Cain and Abel type, too many crucifixions in wooden electric chairs, too many poor and underrepresented people strapped to gurneys waiting for the asp bite of deadly hypodermic needles. While it is true that death offers a type of liberation, few human beings welcome its embrace.

Death frightens us because it stalks the heroic and the cowardly with equal relentlessness. As self-aware beings, we know that our existence is bracketed by birth and death. It is violence that slashes through ordinary lives and grabs our attention. As much as we fear it, we also feel awe and reverence for such a powerful force in the world. Of late, Westerners have become intoxicated with violence. It is the elixir of our generation. We drink deeply, sipping its poisonous nectar, refusing to heed the warnings of past generations that liberation and violence are a volatile mixture.

Do not misunderstand. Because I am a black woman living in the West, I am very familiar with violence, so this is no prissy complaint. Violence lurks in our homes, hides behind the cloak of legality, and wages wars of aggression in the name of freedom. Although we are encouraged to view war games as precursors to peace, we know better. The liberation that we have been seeking is drenched with a peace that surpasses our understanding, and we know deep down in our souls that we have not come close to that goal. Ethicist and attorney William Stringfellow says this about the immediacy of violence in America:

> America is a violent society not because there are not viable alternatives to violence, but because . . . violence has become the first, quick, and often pre-emptive recourse in substitution for the policy or the practice of non-violence.[4]

Stringfellow identifies the ordinariness of violence that is carefully seeded throughout our society. We often presume that it just explodes out of nowhere. We pretend that we could not have possibly seen it coming. Yet, violence has a history that is deeply rooted in the story of human existence. That history is reflected in our systems of justice and religion in very intentional ways. It is difficult to gauge the violent proclivities of a society because there are no meters or scales to measure the ratio of violence to benevolence. However, the punishment that a society metes out to offenders often indicates its justice orientation.

When the initiatives that support an egalitarian society are stalled or when survival for the poor and middle class becomes difficult, the bloodlust for righteous retribution is heightened. Often, however, those systems that imprison and kill, and those factors that sanction or create fertile ground for individual acts of violence, remain hidden from public view. And because they are hidden, we seem to be unaware of our predilections toward exacerbating violence when race, gender, ethnicity, or sexual orientation is involved.

When the recent spate of school shootings began, the nation wondered with nods of distracted amazement that this could happen in predominantly white, middle-class schools. When Latino, Asian, and African American gangs mark their territory with the bodies of their own neighbors, the same information is greeted with resigned acceptance.

Conclusions regarding the inherent criminality of one race or another are reinforced by the statistics regarding high percentages of African Americans in prison. When the first communal response is violence, one hardly considers the horror of its effects with more than passing interest, unless the victims or perpetrators are significantly unique in age or story.

One thing is certain, violence runs deep and wide in our lives. The liberation movements, while peaceful, tapped into that reality. Inevitably, peaceful marchers set off eruptions of hostility that were extraordinary in their brutality. The violent deaths of Oscar Romero, Medgar Evers, Martin Luther King Jr., Malcolm X, Viola Liouzzo, and Emmett Till, to name just a few, shocked us because we saw our own hatred reflected in their broken bodies. When the liberation project was derailed, violence returned to its place as a hidden sociospiritual tributary.

This chapter was unlike the others. It interrupted while I was waiting in court to testify against a defendant who walked into my house and stole property. Life happens even when you are on an assignment as special as this one. My first inclination was to forget about it; it was only property. But in a violent society, when intruders know where you live, you must do something or they'll be back, bearing with them the potential for a violent encounter. Funny part about this is that our helplessness seems complete, for even if you stand your ground and testify, they may still come back.

The wait in court was interminable, and the parade of folks seemed unusually eclectic. Men in muscle shirts sat next to folks caught "driving with a suspended license"; prostitutes stood in orange jumpsuits, looking extremely shopworn. Teens lurched to the podium to testify in baggy pants and a sullen stance. Everyone seemed to be just a hairbreadth away from mayhem, thus all of the milling and watchful bailiffs. As I sat with people I seldom noticed, I wondered if this volatile collective of human beings was anything like a beloved community. I'm serious! Is that idea reserved for the lovely, the well-meaning and the blessed, or is coming together across false boundaries a first step toward the communication that so easily eludes us? I was thinking about this when the conversation topic for this chapter emerged in my thoughts.

Violence is killing us, killing you, killing me, and this conversation was intended to identify and consider options for the survival of besieged people. I would be talking with Ida B. Wells-Barnett, Stanley Tookie Williams, and Huey P. Newton. But the topic was unpleasant enough for me to feel a decided reluctance about the interview. Yet I was hopeful that some of the insights shared might help future generations. I was in the middle of that thought when Barnett appeared.

Ida B. All that effort, and they are still lynching folks.

Writer Well, but *lynching* is such a hard word.

Ida B. The word is no harder than the action, Sister.

Writer I know, but my point is this: although the meaning may be clear to those subjected to its harm, words like *lynching* and *racist* hammer the human brain in ways that cause people to recoil and reject any discussion that might be relevant.

Tookie You may be right, but discomfort with reality should not circumscribe our prophetic proclamation. Those in power may not like the descriptive words, but the wholesale jailing of people of color in disproportionate numbers to the population, and the predominance of their executions, is "lynching" of a different type. But, I'm easy to get along with. If you want to use another word for the systematic destruction of a targeted group of people, that's fine with me. I'm here today because I wanted to meet you, Ma'am [speaking to Wells-Barnett]. You are a bad sister!

I heard that you walked around Memphis with a pistol on each hip. I just had to holler at you. I can't stay long because my transition is recent and I have instruction and life review to complete. But I wanted to add my two cents to this discussion. Where is Huey, we need him to drop some knowledge?

Ida B. Let's just say that I protected myself, just like the time that I bit a street conductor when he tried to make me give up my seat on the trolley for a white passenger, but there was not a violent bone in my body. Violence is always a decision to commit mayhem, and it is just as much of a spiritual entity as grace and mercy. You can't fight violence with more violence because one feeds on the other and both are sustained.

Until Huey gets here, let me school you about the purpose and intent of lynching. Lynching had more to do with dissipating the will of African Americans to resist oppression. Killing is easy. It could have been done without the terror, but the point of terror is the installation of fear in the heart. Once fear is seeded, the option for joy is foreclosed. It was 1889, and I was teaching when I first began writing for the *Free Speech and Headlight*. The paper was owned by Rev. R. Nightingale—the pastor of Beale Street Baptist Church. Because the congregation was large and supportive, the paper became solvent very quickly, and it wasn't long before I could leave my teaching job.[5]

Tookie Once again the church becomes a safe haven for the exercise of rights bestowed by the Constitution and inherited as the legacy of a free Savior. It sounds as if your times were as rough as mine. It was no small thing to offer you support; your pastor must have been a real "dude."

Ida B. Huh?

Tookie You know what I mean—he took a stand and backed you up. He had to be a bit nervous about the whole thing, because after they take someone to a hanging tree, you have to wonder whether or not you will be next.

Ida B. In those type of circumstances, you don't have to wonder; you know that if they can catch you, they will kill you. Except for my public writing, our resistance to oppression had to be done in secret. Public nonviolent resistance would not have worked and instead would have incited a massacre. There seemed to be no way out when I wrote:

> The city of Memphis has demonstrated that neither character nor standing avails the Negro if he dares to protect himself against the white man or become his rival. There is nothing we can do about the lynching now, as we are out-numbered and without arms. The white mob could help itself to ammunition without pay, but the order is rigidly enforced against the selling of guns

to Negroes. There is therefore only one thing left to do; save our money and leave a town which will neither protect our lives and property, nor give us a fair trial in the courts, but takes us out and murders us in cold blood when accused by white persons.[6]

Writer I live in Memphis now, so your struggles have real meaning for me. In any city in the deep South, it is not difficult to conceive of a time when living there was a game of Russian roulette for people of color.

Tookie Violence needs an atmosphere of agreement and is always easier when everyone explicitly or implicitly agrees to it. So you took your own advice and got out of town.

Ida B. It's one thing to sacrifice your own life; it's another thing to put church members and family in peril because you have to tell the truth and tell it in print. My story is well known, but yours doesn't make sense to me. You killed, and then the state killed you. There doesn't seem to be anything redemptive about that. How is the community helped? How are all the young people mired in violence enlightened?

Tookie I may have been a criminal, but I didn't start out to terrorize the neighborhood or the nation. We were being preyed on by gangs of every type. I organized the "hood" to protect the "hood," and before I knew it we were a gang, a more violent and efficient gang. I didn't factor in the rootlessness that had taken hold in our community. The mamas and daddies had to make a choice; they could raise the next generation or work too many jobs for too little money. The pull of the "middle-class lifestyle" won, and the children were relegated to the after-school programs and the streets. And so, the substitute gang families gained power. But by the time they were ready to execute me, I was the rehabilitated founder of the violent Crips street gang. I was a Nobel Prize nominee and the author of children's books that sent a message of hope and escape from a life on the streets. I wasn't even the same person.

Did you know that cells in our body change, so that the person you were at five years old no longer exists when you are twenty? Now, I'm not making excuses—I killed and I was killed—but isn't execution antithetical to the message of rehabilitation that is supposed to undergird the American justice system? When I finally saw the light, they hastened the process to extinguish it. Meantime, I was determined to pour out my life to prepare young hustlers and gangsters to escape the consequences of thug life. And, I don't use that phrase in the way that Tupac Shakur did. He was using the phrase as an inversion of life stratas. He was using the phrase as a call to positive social revolution. When I offered the following statement, I was trying to sound the alarm.

I am confident enough to say to all of you today that if the death penalty is objectively investigated, it will be exposed for what it is—a racist, inhumane and disproportionately allocated system used primarily for poor people. I don't know about any of you, but personally, I can't name one millionaire or billionaire on death row. Can you?

In conclusion, I am fully aware that when I help underprivileged children and speak out against the death penalty, I do so from a vulnerable position. My voice can be silenced by institutional bureaucracy treachery, but the message transcends my life and it is God's will. Meanwhile, while I breathe, I hope.[7]

Writer What does hope mean on death row?

Tookie It means that I finally embraced redemption, and redemption gave me a spiritual platform that shaped my destiny and allowed hope to grow.

My interpretation of redemption differs from the theological. . . . I believe that my redemption symbolizes the end of a bad beginning and a new start. It goes beyond a sense of being liberated from one's sins or atonement in itself. I feel that my redemption mostly or primarily encompasses the ability to reach out to others.[8]

Huey P. And your message lives on, Brother! Talk about dropping knowledge. We'll talk later, Tookie.

Ida B. Peace, Tookie. Let's get together after your classes. Hi, Huey! Now, let me get to the point. For me, lynching is killing by means of unrestrained power. Although they said that lynching was a response to crimes against white women, the statistics prove otherwise.

A partial record of lynching is cited; 285 persons were lynched for causes as follows:

unknown cause 92
no cause 10
prejudice 49
miscegenation 7
informing 12
making threats 11
keeping saloon 3
practicing fraud 5

practicing voodooism 2
bad reputation 8
unpopularity 3
mistaken identity 5
using improper language 3
violation of contract 1
writing insulting letter 2
eloping 2
poisoning horse 1
poisoning well 2
self defense 6
having smallpox 1
asking white woman to marry 1[9]

Talk about the banality of evil—those attending the first NAACP convention in Atlanta (May 8, 1909) were shocked by the serendipity of the causes of lynching. There is this tendency to attribute reason to the most violent and chaotic events. It is a way of bringing those events within the realm of human comprehension. But some things should be required to stand in the daylight of our gaze without the cover of our convenient justifications.

I'm sorry to monopolize so much time, but if you ask me, I don't see much difference in your current prison system. If you consider the plight of minorities in American prisons, you find executions of the guilty and the innocent, but the unbridled power that allows some killers to bond and go free and others to die is frightening. Sociologist Troy Duster suggests that the economic downturn during the 1980s and 1990s led more lower-income youth into the drug trade. Although this era signifies the investment banking boom and the dot.com bust, jobs for unskilled labor tanked and did not rebound.

You didn't need an education or a résumé to sell drugs, but you did have more intense contact with law enforcement and a higher probability of receiving a prison sentence. And so, the prisons began to fill disproportionately with African American men.[10]

Writer I'm sure that the reasons for such high rates of incarceration are complex, but the incarceration statistics are shocking. In the face of such disproportional rates of imprisonment, I want more than civil rights anthems to protect my community. I've always had an admiration for the Deacons and the Black Panthers. The Panthers just celebrated their fortieth anniversary. I remember seeing Erica Huggins (Black Panther Johnny Huggins's widow) on C-SPAN recalling what it was like when the police followed the Black Panthers everywhere they went. She remembers the source of the violence that almost consumed them. She remembers that a police officer had her husband's testicles in his hand and that she was almost nine months pregnant at the time.

So many freedom seekers died labeled criminals when those who grew up during eras of resistance knew that they were just like David's men and Deborah's armies fighting for space to breathe life into a community that was ostensibly free and spiritually bereft and confused. Only in the most functionary and superficial ways could these fighting men and women be called "criminals."

Huey P. You're right, Sis. In North American society, a criminal is a person who breaks laws and is convicted. But African Americans know that the word *criminal* is not a pure identification. It may carry an overlay of color and class despite purportedly neutral applications of justice. This is so much a part of African Diasporan experiential knowledge that subversion of the system and its skewed applications of justice can occur on juries where the issues of race and guilt arise. The majority culture professes to be stunned that inner city juries will free defendants whose guilt seems predetermined by the evidence. However, if the system is killing you, killing me, killing us, then irrespective of "guilt," causing it to malfunction is a revolutionary act.

When Bobby Seale and I started the Panthers, it was for empowerment and self-protection. We were sick of being abused by the police. We saw no way to confront their violence without being armed and ready to protect our people. What we did not anticipate was that our resistance would reach all the way up to J. Edgar Hoover. The shock of seeing us unbend our backs and face them down with guns of our own led to the destruction of the party. In 1968, Hoover ordered the FBI's Counterintelligence Program (COINTELPRO) to begin the organized destabilization of the party. The tools included misinformation, infiltration, surveillance, and assassination.[11]

Writer How were you received by the community?

Huey P. They didn't believe that relief could come, but they appreciated the commitment to community health and welfare. Yes, we carried guns (which is one of the quintessentially American practices), but we also started sickle cell anemia testing centers and lunch programs. Because our neighborhoods were within the confines of the United States, people did not understand that we were prisoners living in territory occupied by the police. They killed at will, but if we dared to raise our heads, they escalated the violence. It occurred to Bobby and me that we lived in a country that believed in violence yet we were supposed to sway and sing about a someday "great getting up morning."

We didn't want promises in the next world; we wanted self-determination, community service, and armed self-defense. If you want to know what we think about liberation, I really can't offer anything new. The ten points of the Panther Party were right for our time and yours.

Ida B. Everyone thinks of Malcolm as the necessary balance for Martin. A violent, lynch-loving nation would not have acceded to those peaceful marches if Malcolm had not stood at the ready with armed and disciplined troops ready to rumble. The Panthers carried on the legacy of both men, serving the community in peaceful and supportive ways and at the same time standing with guns cocked, ready to defend and protect. I don't know why historians try to simplify such a complex legacy into a snippet that relies on violence as the tag. The essence of the idea was found in the ten points developed by the party.

Writer OK, I fail the test. Remind me of the ten points.

Ida B. Let me do it, Huey. I love this platform.

1. We want freedom. We want Power to determine the destiny of our black and oppressed communities.
2. We want full employment for our people.
3. We want an end to the Robbery by the Capitalists of our Black and Oppressed Communities.
4. We want decent housing, fit for the shelter of human beings.
5. We want decent Education for our people that exposes the true nature of this decadent American society. We want education that teaches us our true history and our role in the present day society.
6. We want completely free healthcare for all black and oppressed people.
7. We want an immediate end to police brutality and murder of Black people, other people of color, all oppressed people inside the United States.
8. We want an immediate end to all wars of aggression.
9. We want freedom for all Black and oppressed people now held in U.S. federal, state, county, city and military prisons and jails. We want trials by a jury of peers for all persons charged with so called crimes under the laws of this country.
10. We want land, bread, housing, education, clothing, justice, peace and people's community control of modern technology.[12]

Huey P. Amazing that you remember! In each point, the community heard us say, "We love you and will die to protect you." Even warriors have mothers and children and must feed and care for them. In the ten points, we were asking the impossible from a system that was organized for our destruction. But it was in the articulation that we found our freedom.

As I listened to Ida B. recite the basic tenets of Panther philosophy, I was reminded of all of the righteous women and men who fought for the freedoms we so easily neglect today. There he stood, the legendary Huey P. Newton, a very complex person who understood the need for revolution and at the same time struggled with his own issues of fragmentation.

Huey P. Unless you expect the unexpected, you will be living in a fantasy world that reinforces your belief that you exercise control in the world. Even God surprises us with a temperament that does not fit our preferred divine profile. Don't ever forget that redemption, according to our theology, is steeped in the violence of the cross. It's all part of the mystery. That's why I like the fact that you are trying to keep the cosmos in mind. To be fully human requires an acknowledgment of the mystery of the cosmological context. How can we dare to kill one another when we aren't even certain about the boundaries of life and death? When we execute someone, where do we think that we are sending them? Our theologies are certainly challenged by the death penalty.

Ida B. I've always wondered about that myself. While I was busy fighting to keep black folks from dangling from strong trees, I also wondered about the ultimate goals of these types of executions. Did the lynch mobs think that they were sending these people into the arms of a loving God? The only knowledge that we have of life beyond execution comes in the story of resurrection. The lesson there is that there is life after life and that it is not a punishment. In fact, this may be the ultimate pathway to freedom.

Huey P. Wouldn't it be ironic if the society's desire to destroy engenders liberation for the executed? Now that's a cosmic flip of the script. We were trying to model the antithesis of victim mentality through the ten points. The system wants us to create prisons in our own minds, barbed-wire fences in the communal consciousness so that we don't reach out to help one another, so that we dwell on our victim status instead of imagining and activating our innate freedom.

> All oppressed people must be controlled. Since open force and economic coercion are practical only part of the time, ideology—that is internalized oppression, the voice in the head—is brought in to fill up the gap. When people discover their own power, governments tremble. Therefore, in addition to all the other things

that are done to control people, their own strength must be made taboo to them.[13]

Sister Russ is talking about the ways in which women are controlled by an ethos that invites them to build their own cages. Even when power is obtained, it must be filtered through an internal reality that, unbeknownst to the assertive external creature, is cowering and laden with fetid and familiar myths of inferiority. People of color often live divided lives, believing that they can "overcome" and at the same time battling internalized bastions of inferiority.

Writer This syndrome is most evident among those for whom striving is the matrix of their lives. It is as if they must keep moving to avoid the silences that would confront their inner chaos. There is a universe within. For some, the universe is an expanding and challenging reality, but like any good galaxy, there are also black holes into which everything is sucked into oblivion.

Ida B. Well, that's a happy thought. Look, I'm more optimistic. There is always more than one way to reach a goal, and I think that Huey mapped the road to a liberation that can be sustained through the Panther manifesto. Liberation begins in the consciousness and in the spirit of the people. By the time people are marching in the streets or taking up arms, it has already flowered. Once planted, it is a seed that cannot be dislodged. Often we don't see the fruit of generations that tended, weeded, and waited for the harvest. For each generation, the time to obtain liberation is always "now."

Writer Yes, but the word *now* has an immediacy that seems to be connected to a view of the universe that is objective and solid. The new cosmology tells us something different. I'm just wondering what would happen if we addressed the violence of the human community from the perspective of nonlocality. Humor me for a second. Each of the scientific theories that were addressed in previous conversations connected well to the theme.

Huey P. The Panther manifesto was aimed at restoration (self-determination), redress (protection against police brutality), and remedies (community survival). The ideas represented presumed a certain Newtonian order to the world. But suppose the world is not made up of oppositional forces; suppose everything is connected to everything else. What then?

Writer At this point, let me summarize Alain Aspect's research on connections and nonlocality.

In 1982, a remarkable event took place. At the University of Paris a research team led by physicist Alain Aspect performed what may turn out to be one of the most important experiments of the 20th century. . . . Aspect and his team discovered that under certain circumstances subatomic particles such as electrons are able to instantaneously communicate with each other regardless of the distance separating them. It doesn't matter whether they are 10 feet or 10 billion miles apart. Somehow each particle always seems to know what the other is doing.[14]

There are other scientific experiments that point toward the connection of all living things, a sort of quantum entanglement.

Huey P. If this is true, then the evil that we do individually is never a solitary act of antisocial behavior; rather, it is the first volley that sends a million pins crashing into one another. When we wage wars, we are fighting the unity that is the reality of the universe and harming ourselves. I think that I understand this.

Ida B. Although they didn't realize it, the lynch mobs were tying a cosmic noose around their own necks and the future of their children, because the fact that we stand someone up against a firing squad or inject them with lethal drugs does not disconnect us from one another. I realize that these interconnections are uncomfortable, to say the least, because there are just some folks that you don't want in the family.

I may say that I love all people, but when you produce the enemies from the past and the present, I don't know that I am willing to claim a bond to the Grand Wizard of the Klan or Hitler or even my nosy neighbors. This oneness thing requires that we realize that liberation is not from someone or some system. Liberation is a new future that we must create together, learning from all of the past wrongs and moving toward an uncertain future.

Huey P. Perhaps that is why the discourse of reconciliation is so locked in black/white discourse. It is as if the battle can't be disengaged. Once connected through violence and abuse, I have to wonder whether it is ever possible to go our separate ways, if separation is even a possibility in this reality. And this takes us back to the prisons. We are connected to the ones that we execute and imprison in ways that contribute to our own continuing social ills.

Even without knowing any science, we know that extremely punitive systems don't contribute to healthy social change. If society was to suddenly become the site of blessings and well-being because of the removal and execution of offenders, we would be silenced by the facts. But if anything, use of the state as an arm of violence seems to make the violence in society worse.

Ida B. Everything that we are doing is based on the belief that we are separate human beings. If we began to see ourselves connected to every life that we come into contact with, how would we go on? There is a scripture that says that you work out your own salvation with fear and trembling.

Huey P. Suppose the process of working it out requires that we realize deep connections because, spiritually and actually, it is impossible to just save ourselves.

Ida B. Perhaps our studied indifference toward one another is a way of masking our fear of deep connections. Here is the amazing part. Even without knowing about nonlocal effects and interconnections, your Panther manifesto is a request to recognize our relationships and to stop the acts that support the delusion of separation. I can't continue without asking you something. How did you go from being an icon of revolution to being killed by a drug dealer on the same streets that you tried to redeem?

Huey P. If this interconnection stuff is right, then it should sustain you when you are down and lift you when everything around you seems to fall away. We needed each other so much. We were struggling against powers and principalities, and we needed to know that our friends were with us. When the disinformation and counterintelligence project began, they destroyed our trust, our hope, and our will to persevere. From the distance of historical reflection, it looks as if we could have overcome their diligence, but it was not possible. If we had known, it would have been different. Instead, we had to serve and protect while steeped in the acrid reality of betrayal. Moreover, the struggle for liberation was so intense that the spiritual and emotional energy needed to continue sometimes blocked the connections needed to survive.

I also believe that there are only so many killings that we can absorb in the body and the spirit without relinquishing the tenuous hold that we have on our own lives. I died when Johnny Huggins died; I died when Fred Hampton was killed. I was gravely wounded during shootouts that didn't involve me. I began to fade even as I was fighting for the survival of my people. I don't want to glamorize my end. There had to be a better way. I always longed for a more noble end, but it was what it was. Sometimes the most beloved leaders face an ignominious end (Christ was one).

I don't compare myself to Christ, although perhaps I should if I am supposed to be a living image of the one who died as a criminal. This is a fragment of a poem that I wrote:"I heard god call, I got my gun and waited. When [god] appeared, I realized and took the gun from my head."[15] I can't deny who I was because to do so makes the lessons that I learned null and void. Another poem fragment: "if I define myself as my body, I deny myself my universe, I diminish myself."[16] In the end, I was of the people and died addicted and estranged, like many of my people, but always with a readiness for the next struggle for justice in my spirit.

Ida B. I don't think that any of us is happy with our final moments. There is always such a desire to complete what we started. I died up North, far from the battleground of the South. In a way, I felt estranged from the struggle for justice. Listen, Huey, we were not superheroes; we were ordinary people with failings and gifts. We did what we could, and now science is backing us up. We are learning that if there is to be a future, it will be a future of intertwined lives.

Huey P. If I had to leave a final word with future generations, it would be that every liberation movement must be tied to more than the fact of abuse and oppression, must exceed the political and social boundaries of the immediate circumstances, and must have in its quest for transformation at least some element that invites enemies and oppressors to enter into the process. We considered King to be too middle class and too soft on self-defense and community empowerment, but his coalition always included those who crossed the lines of race and social status to join forces with a righteous cause.

Ida B. The liberation that we seek is already here. It is in the lives and imaginations of activists and those trying to maintain the status quo. We can live out of the myth of our separateness or make visible the quantum and spiritual entanglements that make violence toward one another an act of mutual destruction.

Summary: What Is the Relationship between Violence and Liberation?
Writer I'm having less and less to say after each interview. It may be the accumulation of wisdom. It helps to include within the notion of violence a systematic and state-sponsored approach. Most of the conversation around state executions focuses on debates about morality, character, and shared values.

Sarah It looks as if you are headed toward the "bleeding heart liberal" school of analysis. I can hear the challenge posed to you right now: "Would you feel the same way if they killed your family member?" Would you still be willing to entertain a systems approach to the issue?

Writer If the personal is political, then I would have no choice. When the jails rehabilitate, when mental health service providers are available to all in need, when law enforcement agencies require officers to respect the humanity of all persons, when poverty and discrimination are no longer providing structural support for the "powers that be," then we can talk about individual feelings about individual acts.

Sarah You raise important points. If violence has consequences in an interconnected universe, then liberation requires humankind to relinquish its addiction to punishment, retribution, and war.

Writer Violence in any form is killing you, me, and us, and yet the universe is a place of violent explosions, dying stars, and crashing asteroids, to name just a few of the more catastrophic events.

Sarah This is one discussion that can't be tied up neatly. I think that there is a difference between death-dealing violence and the creative genesis of the universe that sometimes leads to the destruction of life forces but also nurtures regeneration. The question is whether we will spend our time lamenting the violence endemic to human communities or looking for "leading causes of life" in everyone.[17]

The Exercises

1. Have you been a victim of crime? Reflect on the event and its meaning. Are there elements that transcend the factual elements of the event?

2. In one paragraph, describe a crime that is well known nationally or locally. List five words that describe your feelings about the crime. Then write words that mean the exact opposite of the five that you have chosen (e.g., hate/love). Be ready to explain how both words are part of the story.

3. Describe a cosmological alternative to the death penalty. Think in terms of wide-ranging effects and creative responses to redemption and reconciliation.

4. Ida B. Wells-Barnett was an investigative journalist. Choose a topic related to violence in society, and write a one-paragraph editorial on the topic.

5. Would nonviolence have worked for Huey P. Newton or Ida B. Wells-Barnett in their contexts? List three reasons to support your position.

For Further Reflection

Ida B. Wells-Barnett
Baker, Lee D. "Ida B. Wells-Barnett and Her Passion for Justice." Available at http://www.duke.edu/~ldbaker/classes/AAIH/caaih/ibwells/ibwbk-grd.html (accessed June 8, 2008).

Holt, Thomas "The Lonely Warrior: Ida B. Wells-Barnett." In *Black Leaders of the Twentieth Century*, ed. John Hope Franklin and August Meier. Urbana: University of Illinois Press, 1982.

"Ida B. Wells-Barnett: Postscript by W.E.B. Du Bois." *Crisis* (June 1931): 207.

Wells-Barnett, Ida B. *Crusade for Justice: The Autobiography of Ida B. Wells*. Ed. Alfreda B. Duster [Wells-Barnett's daughter]. Chicago: University of Chicago Press, 1970.

———. Diary entry, January 13, 1930. In *The Memphis Diary of Ida B. Wells: An Intimate Portrait of the Activist as a Young Woman,* ed. Miriam DeCosta Willis, 168. Boston: Beacon, 1995.

———. *On Lynchings: Southern Horrors, a Red Record, Mob Rule in New Orleans*. New York: Arno, 1969.

Wells-Barnett, Ida B., and Trudier Harris-Lopez. *Selected Works of Ida B. Wells-Barnett*. New York: Oxford University Press, 1991.

For a few recent examples of analyses of lynching, see W. Fitzhugh Brundage, *Under Sentence of Death: Lynching in the South* (Chapel Hill: University of North Carolina Press, 1997); Stewart Emory Tolnay, *A Festival of Violence: An Analysis of Southern Lynchings, 1882–1930* (Urbana: University of Illinois Press, 1995); and Crystal Nicole Feimster, "Ladies and Lynching: The Gendered Discourse of Mob Violence in the New South, 1880–1930" (Ph.D. diss., Princeton University, 2000).

Stanley Tookie Williams
Pamphlets, Reading Level 9–12 (with Barbara Cottman Becnel)

———. *Gangs and Drugs* (Stanley Tookie Williams Speaks Out against Gang Violence). 1997.

———. *Gangs and Self Esteem: Tookie Speaks Out against Gang Violence*. 1999.

———. *Gangs and the Abuse of Power* (Stanley Tookie Williams Speaks Out against Gangs). 1997.

———. *Gangs and Wanting to Belong* (Stanley Tookie Williams Speaks Out against Gang Violence). 1997.

Pamphlets, Reading Level 4–8
Williams, Stanley Tookie. *Gangs and Your Neighborhood* (Stanley Tookie Williams Speaks Out against Gang Violence). 1997.
———. *Life in Prison.* (Royalties donated to the Institute for the Prevention of Youth Violence). 1998.

Books
Williams, Stanley Tookie. *Redemption: From Original Gangster to Nobel Prize Nominee—The Extraordinary Life Story of Stanley Tookie Williams.* Lytham, U. K.: Milo, 2004.
Williams, Stanley Tookie, and Barbara Cottman Becnel. *Blue Rage, Black Redemption: A Memoir.* New York: Touchstone, 2005.

Huey P. Newton
Churchill, Ward, and Jim Vander Wall. *Agents of Repression: The FBI's Secret Wars against the Black Panther Party and the American Indian Movement.* Boston: South End, 1990.
Erikson, Erik H., and Huey P. Newton. *In Search of Common Ground: Conversations with Erik H. Erikson and Huey P. Newton.* New York: Norton, 1973.
Heath, G. Louis. *The Black Panther Leaders Speak: Huey P. Newton, Bobby Seale, Eldridge Cleaver and Company Speak Out through the Black Panther Party's Official Newspaper.* Metuchen, N.J.: Scarecrow, 1976.
Jeffries, Judson L. *Huey P. Newton: The Radical Theorist.* Jackson: University Press of Mississippi, 2002.
Newton, Huey P. *The Huey P. Newton Reader.* Ed. David Hilliard and Donald Weise. New York: Seven Stories, 2002.
Pearson, Hugh. *The Shadow of the Panther: Huey Newton and the Price of Black Power in America.* Reading, Mass.: Addison-Wesley, 1994.
Seale, Bobby. *Seize the Time: The Story of the Black Panther Party and Huey P. Newton.* Baltimore: Black Classic, 1991.

With a camera you can show things that you like about the universe, things that you hate about the universe: it is capable of doing both.
—Gordon Parks

8. Liberation and the Art of Creative Imagination: Prophetic Expression

My music is not for everyone. It's only for the strong-willed, the [street] soldiers music. It's not like party music—I mean, you could gig to it, but it's spiritual. My music is spiritual. It's like Negro spirituals, except for the fact that I'm not saying "We shall Overcome." I'm saying that we are overcome.
—Tupac Shakur

Tupac Shakur, Phillis Wheatley, Langston Hughes, and Gwendolyn Brooks share insights regarding the role of art and liberation. The artists challenge the status quo and consider the power of creativity to chart a new course toward liberation.

The Elders

Tupac Amaru Shakur, 1971–1996

I am not saying I'm going to change the world. But I guarantee that I will spark the brain that will change the world.[1]

Tupac Amaru (Quechan/Incan for "Shining Serpent") Shakur (Arabic for "Thankful for God"), artist extraordinaire and son of two Black Panther parents, is larger than life even in death. His raps reflect the frustration, grit, and creativity of life in the "hood." He rapped about reality in language that was raw, poetic, and sometimes misogynist. His descriptions of drug use and violence make most folks cringe, and yet there is something playful and prophetic about many of his oracles. I am using the first person to describe Pac because his death at age twenty-five seems less real than the sound of his voice still chanting "I Ain't Mad at Ya" on the radio twelve years later. His song "Dear Mama" is the most honest and touching tribute to an imperfect but beloved parent.

He was born in New York and lived in California, but that information does not disclose much about the man. He lived fast and skirted the edges of safety and risk, on one hand wanting to develop a "thug life" philosophy that would help to liberate kids in the "hood" from their difficult circumstances and on the other hand allowing artistic depictions of gangster swagger to morph into a deadly drama. When East Coast/West Coast tempers flared and the assassination of rappers followed, Tupac was gunned down in Las Vegas in 1996.

Phillis Wheatley, 1753 (or 55)–1784

God has implanted a principle, which we call love of freedom; it is impatient of oppression, and pants for deliverance—and by the leave of our modern Egyptians I will assert that the same principle lives in us. God grant deliverance in his [sic] own way and time, and get him [sic] honour upon all those whose avarice impels them to countenance and help forward the calamities of their fellow creatures. This I desire not for their hurt, but to convince them of the strange absurdity of their conduct whose words and actions are so diametrically opposite. How well the cry for liberty, and the reverse disposition for the exercise of oppressive power over others agree, I humbly think it does not require the penetration of a philosopher to determine.[2]

To the ears of a culture steeped in hip-hop, rock, and spoken word poetry, the classical rhythms of an eighteenth-century poet seem odd. But even more unusual is the fact that these words were penned by an African woman enslaved in Boston, Massachusetts. Historians allude to a genteel bondage, but it was bondage nonetheless. It is presumed that Phillis Wheatley was captured in Senegal-Gambia and sold at age eight to the Wheatley family of Boston. Phillis was named after the slave ship that brought her to New England (the *Phillis*). Her last name was the name of her owners. She was a stolen child with a borrowed identity, caught between realms.

Taught to read and write, she began writing poetry at a time when the prevailing opinion in Europe and the United States was that Africans were subhuman. Her poetry was deemed by abolitionists to be a prime example of the intellectual potential of Africans in the Diaspora. She traveled to Europe, published a poetry collection, and upon the death of her slave owners was freed. She married and struggled with her husband to sustain a free life in a segregated society. Some wonder why her language was not more revolutionary; others wonder if she actually authored the poems. We can only guess as to the motivation and thought processes that inspired her words. She may not fit the postmodern mold of a revolutionary wordsmith, but she is the first published female poet of African descent to be recognized during her era.

(James) Langston Hughes, 1902–1967

We younger Negro artists now intend to express our individual darkskinned selves without fear or shame. If white people are pleased we are glad. If they aren't, it doesn't matter. We know we are beautiful and ugly too. If colored people are pleased we are glad. If they are not, their displeasure doesn't matter either. We build our temples for tomorrow, as strong as we know how and we stand on the top of the mountain, free within ourselves.[3]

Langston Hughes seems to have been born to write. His biographers are careful to make mention of a family history that included politicians and activists. No one would have been surprised if he had followed that route, but instead, by the eighth grade, he was writing poems and being recognized for his talent. A short stint in college with an engineering major ended when Hughes decided to see where his artistic endeavors would take him. He traveled to West Africa and Europe, received a degree from Lincoln University, and won numerous awards. He was a prolific writer and flourished during the Harlem Renaissance. He encouraged African American artists to speak in their own voices and to avoid the mimicry of the dominant culture that might increase the probability of acceptance but would inevitably diminish their prophetic artistic expressions.

Gwendolyn Brooks, 1917–2000

Art hurts. Art urges voyages—and it is easier to stay at home.[4]

Gwendolyn Brooks was shy and spent her childhood writing and reflecting on life. Reading the numerous biographies about her, it becomes clear that her shyness made it difficult to relate to her peers. She was born in Topeka, Kansas, but lived most of her life in Chicago. Brooks married and continued to write. She became the first African American to win a Nobel Prize for Literature for a collection of poems. In James Weldon Johnson and Langston Hughes, she found kindred spirits and a community of writers who would challenge her and urge her toward the incisiveness that marked her later work. Her writing style was intellectual and meticulous, with a careful economy of words. During the Black Arts movement of the 1960s, Brooks was influenced by the revolutionary discourse that had always been a subtext in her work. Brooks loved words, and they served her well as she articulated the angst of a generation.

The Conversation

If art has the power to liberate, it must surpass its time. That's how you know that you really have something. It is not while the madness is upon you, when you are thinking of cutting off your ears or lying upside down in the Sistine chapel or exchanging gunfire after a street dance concert. It is after you are gone and the art that you dedicated your life to carries on without you. It is the music of a deaf Beethoven that the universe could hear, and Tupac's musical blessing of his "Dear Mama," who despite addiction loved him into his art. Art for those who are marginalized is never neat. There are no classes or nudes to draw, just a moment in time when you compose or dance or paint yourself free, even if it is only on a subway wall.

If liberation has a cosmological configuration, then we are closest to it in the arts. For the most part, we don't live free; we do what is expected and make the choices that allow us to conform. Most of us have been trained into submission, educated toward political correctness, churched toward "decency and order." Soon there are no places left to be really free. Inevitably, when the inclination to "be free" escapes the confines of social expectations, we get nervous. We like to talk free, sing free, and sometimes march free, but to live as if there is a God of liberation is another thing entirely. To allow the rhythm of Cab Calloway's high step or Marian Anderson's arias to infuse our imagination is an opportunity to open ourselves to a form of liberation that transcends the everyday.

Freedom is hard to recognize if you haven't lived it. Sometimes it is just a different rhythm, a truth-telling rap, or a painting of a tomato soup can. Freedom expressed through the arts is not static; it dares its audience to observe and then to join in, to sing the chorus of its unchained melody, to try out the dance steps or mimic the artist's independence. The arts invite the observer to see what liberation looks like in a poem or on a stage and to revisit the experience again and again. Eventually, we have the opportunity to imprint the artist's creative courage on our own spirits.

I knew who was supposed to show up for the meeting, but who knew which spirits would actually attend this gathering of artists. The odd part of this interview would be the strange betwixt and between realm that these artists occupied. They were part of a "beyond the veil" conversation, but they were not "dead" to my generation. People such as Gordon Parks, Lorraine Hansberry, and Langston Hughes poured their lives into the arts. An artist bequeaths poems and plays, watercolors and black-and-white photos, and as a consequence, that which they so freely gave during life continues with dynamism when they transition. Every time the curtain goes up on a Lorraine Hansberry play, she lives. It's the same with Langston.

Langston Sure ya right. I was life when I lived. Why wouldn't I be life in death?

No zoot suit for Langston. He seemed to be in a loose-fitting, black, monochromatic dashiki with loose-flowing pants. I didn't know what to say because I was still in the midst of my own reverie, but I gathered my courage.

Writer Are you one of the interviewees, Sir?"

Langston If I'm not, I'm going to Pac the place up!

Tupac Now wait a minute, Brother, let's keep it real. I played the game and then the game played me, but to tell future generations that violence and my stuff is synonymous is whack.

Writer OK, I know that I'm supposed to be quiet, but I can't. Did my favorite rapper just use the word *whack*?

Tupac Sure did. I'd do a "golly gee" if it would help the folks to understand my idea of liberation. For me, it was about taking the denigration and flippin' it. Now that's free. It was the step beyond the James Brown move. He did "I'm black and I'm proud" at a time when calling someone black was a fighting word, when class distinctions between the homies from the hood and the upwardly mobile college kids were clear and lethal.

I had no real impression of Pac's embodiment, although I sensed the same strength and intention that he had lived.

Tupac That's because my transition is recent. I'm still reviewing and understanding the why and the wherefore of the violence and the power of the words. Because you still remember what I looked like in life, there is no necessity to "present." Now, Langston, that old "G," needs to present something with his antique a—.

Langston I've got your antique . . .

They were both laughing and playing and trading rhymes, and talking at the same time, when another voice intervened. It was a formal and proper-sounding woman.

Langston It's Phillis. Are we going back that far?

Wheatley You really are new kids here. There is no back or front, no young or old, just grace and light and insight. I'm here because this was becoming a men's club and the women have something to say about art and liberation.

Tupac Mad respect, Sister, you were writing when society wasn't sure that blacks were human, but I do have a question for you. What was up with the dissing of Africa in your work? I know that you had a Christian thing going, but did you really think that becoming a Christian required immediate hatred of self and the motherland? You don't mind if I quote you?

> Twas mercy brought me from my *Pagan* land,
> Taught my benighted soul to understand
> That there's a God, that there's *a Saviour* too:
> Once I redemption neither fought now knew,
> Some view our sable race with scornful eye,
> Their colour is a diabolic die.
> Remember, *Christians, Negroes,* black as *Cain,*
> May be refined, and join th' angelic train.[5]

Wheatley For my time, this was revolutionary, although I admit that images of Africa as pagan and savage reflect the same type of myopia that all of us occasionally slip into. You express the same thing in your poetry. Don't you say something about having to use five shots to kill someone, because N.'s are hard to kill on your block? Unless I miss my guess, transition has probably changed that view.

Tupac Not really, and I'm not just playing tough, Sister. Y'all are going to be talking about the cosmological implications of liberation. Well, the bangers and rappers were only a microcosm of an infected whole. Like a hologram, which includes the entire entity in each small piece, rappers reflected the societies that we lived in.

I'm not trying to justify the vernacular, or the amplified "dozens," or the verbal battling that could be found in *any* barbershop. We put it to a beat. My point is that the beefin' and banging that I rapped about was small stuff compared to what the political and military gangstas accomplished when they shanghied elections, started wars for their private purposes, and abandoned the poor and the middle class in the land of the free and the home of the brave. Who is more deluded and dangerous—a black teen rhyming about the sex and drugs at his door or the powers that be, spending billions of dollars a day for war and none for national health care? There are least two genocides that went down in the world while people discussed "disrespect" and worried about whether rap caused violence. Something is twisted.

Wheatley Something was twisted in my era also, but I understand what you mean about an overarching society that values conquest, war, and a rigid class structure that keeps most people outside of the bounty of one of the richest countries in the world. You rapped about your abuse at the hands of that society, and I raised my voice in the midst of mine. If you have been silenced and abused, it doesn't really matter what you do with your voice when you find it. It would be better to sing a pretty song, but yelling is also just fine.

The point is to say something real from a place where there is only anguish. I was kidnapped and brought to America from Senegal. I don't remember much about my home except that I loved it. When I speak of love of Africa, it is not with the detail of one who can recall geography or specific memories of daily life. I loved my homeland because it symbolizes freedom for me. This home was free, and before I was taken, it had already stamped its love of liberty on my soul.

When I became a Christian, it felt as if I had become a citizen of a new land. It is this entry into God's spiritual nation that I compared with my homeland. I understand that nowadays Christians join churches and then go about their own business as usual. For us, becoming a Christian meant that we moved our hearts and souls into the center of God's liberative reality and tried to dwell there always. Nothing compares, not homeland, not personal relationships, nothing. I can't defend the heart of a poet, but this is my thinking.

The last thing that I want to say about my reputation is that your critique is not new, Mr. Shakur. The criticism of my work reached its height in the sixties. Henry Louis Gates said this:

> That Phillis Wheatley is not a household word within the black community is owing largely to one poem that she wrote, an

eight line poem entitled "On Being Brought from Africa to America."

. . . [This] has been the most reviled poem in African American literature. To speak in such glowing terms about the "mercy" manifested by the slave trade was not exactly going to endear Miss Wheatley to black power advocates of the 1960's.[6]

Tupac I can't disagree with the brother. Is there any explanation?

Wheatley None, except that context is everything. I am in such a different state of mind now that I can't speak from anything but spirit memory. To think that an enslaved child would revere her captives is absurd. But I am intrigued by an explanation that Henry Louis Gates included in a recent address. Gates reports that he received a fax from a freelance writer who studied the eight-line poem that has caused such controversy, "On Being Brought from Africa to America," and has decided that it is an anagram. When you rearrange the letters in the poem, you get the following plea.

> Hail Brethren in Christ! Have ye forgotten God's word?
> Scriptures teach us that bondage is wrong. His own
> greedy kin sold Joseph into slavery. Is there no balm
> in Gilead? God made us all. Aren't African men [and
> women] born to be free? So am I. Ye commit so brute
> a crime on us. But we can change thy attitude, America
> manumit our race, I thank the Lord.[7]

Tupac It sounds like someone had too much time on their hands if they were able to figure this out. So you protested, but in a hidden way? Give me a moment to absorb what you just said, Sister.

Brooks Look, I can't stay. My transition is also recent, and I have homework to do, but you are letting Pac off the hook. Pac is holding Phillis accountable for her words. Well, what about the women and the misogyny and the glamorization of violence and drugs?

Brooks was breathless but pointed.

Tupac Aww, come on . . . give me a break, Gwendolyn. I wrote songs that expressed love for my mother, wondered about the black Jesus, told black women to "keep their heads up." Don't y'all remember the tradition of verbal jousting, "the dozens" and storytelling in the tradition of the griots? Not one of us believed that the wives, mothers, sisters, and daughters were "ho's" any more than those of us running the streets were soldiers. I'm trying to make sense out of a short lifetime packed with contrasting and conflicting events.

I wrote poetry and used a writer's perspective to tell my story. But what I didn't realize was that it wasn't my story at all that I was telling. It was the story of all people going through something. Real liberation comes with the realization that connections to one another have less to do with the apparent alliances based on gender, race, and neighborhood gang than they do with shared understandings about the ways of the world. But we don't have to agree with one another to find freedom together.

Must I make music that sounds like a Hallmark greeting card to serve the community? We are told to be role models, but don't you have to hang around to model anything for the next generation? The black middle class abandoned the ghetto for better digs in the suburbs. When they left, the neighbor role models left with them. So we chose larger-than-life entertainers to mimic. It doesn't take a lot of money to follow a neighbor's example, but sports and rap stars are another matter entirely. Then the drugs came. They were brought into the inner city. We didn't own jets, but somehow massive supplies were getting to my front door from far-flung places, through customs and all the way to L.A.

My mom didn't go to Afghanistan or Colombia, South America, to get addicted; someone brought it to her door. There was an unending supply that left our women few choices to keep the habit going. What I do regret is that I didn't spend enough time exposing the system behind the observable situation. We rapped about the context of our lives, what we saw and what we lived. There was so much more, but not enough time.

Brooks I think that in some ways your generation became captive to the market economy and its appropriation of your craft, but I agree that the truth-telling discourse and that "in your face" style that tossed politeness out the window changed the way that we think about the "hood." I still have some problems with the language about women, but I truly believe that the art of rap freed a generation.

Kids who had to buy clothes two sizes too big so that they would last until they grew into them turned it into a style that everyone wanted to wear. The life that they lived became the art that changed the rhetoric of a generation. But the other side of that coin is that I don't think that those of you who rapped freely were free indeed. You got absorbed into the corporate money game, and that changed the stakes and the discourse. Suddenly, you were being pushed to live the lives that you rapped about in ways that could not be sustained.

Langston But even with the East and West Coast battle, and the deaths, the raps continue because they are familiar rhythms that emerge in every generation. Even my poems connected to that primal cadence.

Tupac Props, Brother, you were saying what could not be said. You had this thing about women in and out of the house that was hardcore, in soft verse. It was real. You were about something.

Brooks I'm not certain that you're getting my point. We are quibbling about small things when we should be talking about the energy and depth that emerge from the words. Think about the DNA letter sequencing. The letters form potential words in the very depths of your being. There is a word within each of us, and we have to say what we hear. Now, when I hear you, Pac, I may not hear you at all, because my own word is dancing in *my* head.

Langston That DNA stuff is like unfolding mystery. Just suppose that we are born with messages from God in our innermost being and that we have the option to live out of that unfolding internal message. I can't think of anything more cosmic than this.

Wheatley The cosmic event for artists is that we survive at all and that we choose to live out of a reality that stands against the values of wealth as an ultimate goal, because we keep the craft going even when we can't get paid to save our lives. I think that I understand the point that Pac was making when he "checked" me on my view of Africa, He wondered why my language was so accommodating in the midst of slavery. My only response is that I wrote what bubbled up from within. There is no litmus test. Art is the essence of freedom.

Tupac Okay, Sis, did you just say "checked"? I don't want to beef with you; I just want to reclaim the play in the art.

Wheatley Pac, did you just say "reclaimed"?

Tupac In the written art form, there is always a sense of play, with words, with one another, that becomes almost intoxicating. We would go into the studio, and we would be drunk on words . . . for real. I mean just "saying" in a way that felt as if it was coming through, not out of, us. Even the rivalry stuff was play. We weren't dropping bombs on anybody. We were warring with words, until bullets started to kill the joy of the joust. In Africa, in South America, folks challenge and dare with words. It's play, but on a cosmic level because it inverts the game.

Langston That's what the Harlem Renaissance had at its core . . . play. It was time to breathe, to let the struggles of oppression rest while we let the inner expressions of dance and song and words have their way with us. We had been picking cotton, and dying with dignity in the most undignified of circumstances, and making eloquent speeches (God bless Frederick Douglass and his role-model soul). We decided without deciding that there was no point in trying to convince America that we were human and deserving of respect. It was time for relief and joy.

Wheatley In the spirit of Frederick Douglass, I'm going to interject a serious note.

We are talking about our own development as writers and what makes us free to do our work, but what is our responsibility to the community, to its members' liberation from the constraints to present themselves as carbon copies of the majority culture? Are we supposed to make our people look good to the entire world through our art, or is it art for art's sake? Has anyone resolved the W. E. B. DuBois–Alain Locke debate?

I saw them just the other day; they were talking with James Brown and Lorraine Hansberry about pride and giftedness in future generations. Although DuBois is better known for his social and political theory, he is also the "theoretical father of many of the militant writers of the 1960's and '70's."[8] DuBois believed that art should be one of the primary tools in the reconstruction of images of African Diasporan humanity. To accomplish this purpose, he wrote a pageant in 1911 entitled *The Star of Ethiopia* and developed two theatre companies.[9]

Tupac And if I remember my history, Alain Locke was the first Rhodes scholar. He wanted art to tell it like it is . . . to depict the problems and issues of the inner life of African Americans with foibles and problems included. DuBois wanted to present the noble side of African American life.

Langston The debate encapsulates the problems that racism spawns.. Back in the day, we had to choose between expressions of our fallible and wildly interesting humanity or carefully formed symbolic images that made the case that we were also human.

> We are so used to seeing the truth distorted to our despite, that whenever we are portrayed on canvas, in story or on the stage, as simple humans with human frailties, we rebel. We want everything said about us to tell of the best and highest and noblest in us. We insist that our Art and Propaganda be one. This is wrong and the end is harmful. When the artist paints us, s/he has the right to paint us whole and not ignore everything which is not as perfect as we would wish it to be.[10]

The debate continues. We don't know whether to let our art speak or to give it a script that testifies to our moral character as a people. Isn't that the very mark of inferiority, trying through our art to plead with the dominant culture for respect? Locke captured that thought.

> It perpetuates the position of group inferiority even in crying out against it. For it lives and speaks under the shadow of a dominant majority whom it harangues, cajoles, threatens or supplicates.[11]

Tupac So let me get this straight. The argument that rap is embarrassing the race is just another version of the debate between W. E. B. DuBois and Alain Locke? So, when the haters are blaming rappers for bringing the race down, destroying the children's minds, and so forth, they are engaging the DuBois part of the debate. They want us to rap some version of "a mind is a terrible thing to waste." Some of us did that, but others of us felt as if we had to describe the smell of a urine-soaked project hallway. Right or wrong, to let the powerless be heard is a first step toward freedom.

We are supposed to be talking about the relationship between art and liberation. How can we be free if we can't even tell the truth about what's going on down in the "hood"? Most of us did not live in the fantasy world that television depicts. Can anyone besides me see that the rappers freed the black middle class and the white youth from their own self-delusion? We said what couldn't be said in proper company.

We exposed the dissolution of the black community, and when the black church had no real response to the truth that we were spitting, they also stood exposed. We fired the imagination of young white teens alienated from their families and from people of color all over the world. And all anybody can talk about is whether we called someone a "ho." Yes, we were captive to a lot of the nonsense, but we didn't create it.

Langston You're right, Pac. I mess with you, Brother, but I think of your generation as the Joshua gang. Moses and the civil rights generation have walked and marched the people through the desert; now they turn to the young turks to give them some insight as to the lay of the promised land. You come back to us telling us real stories, breaking through the hypnotism of a society that has fallen in love with its own mythologies. It doesn't matter that your clothes are too big and your grills are blingin'; you are telling us that there are giants in the land.

This is more than just a "for or against" argument about the rhetoric of rap; it's also about history. I don't see how any liberation can stick if the people who seek its elusive embrace are ahistorical. The universe encodes within it a history of beginnings and endings, and the stories that the ancestors bring bear witness to your history.

Brooks Art offers many opportunities for liberation. I found my voice through poetry, but when we express ourselves through the arts, we inevitably tap into the taken-for-granted legacy of an African worldview. I like what ethicist Peter Paris says.

> African arts are to enhance the everyday life of the people, not primarily to change their conditions but to enable the people to see and hear and feel beauty. As long as the people enjoy beauty, they do not succumb to the tragic elements in their midst. Their spirits are uplifted, and in that way the arts preserve and promote the wellbeing of the community.[12]

I really understand this.

Tupac A community that is intact and supportive is both powerful and beautiful.

Writer [It was time to interject.] Can we talk about a few specific things? Can we talk about art and cosmology, and the leadership that artists can provide to besieged communities?

I paused because I suddenly became aware that the voices were primarily from writers. Was my own bias as a writer filtering out artists from other disciplines? I was learning that my own interests always came into play. There was no neutral ground.

Langston Well, I'm the elder statesperson around here, so let me start this riff. The contribution from the arts toward a cosmological view of liberation comes in its ability to cross boundaries, to expose the hidden, and to inspire. We won't be free in our own little corrals; it takes a more extreme aesthetic. A cosmological view of liberation is first projected on stages and through music, dance, and art that flow out of the prophetic imagination of the artist. Long before the legal battles began and the marches were launched for the liberation of black people, we imagined ourselves free through spirituals and southern blues.

Tupac Also, what I've learned since I've been here is that there is a rhythm in the universe that is the heart of wisdom. When we stop trying to battle with our weapons and tap into the resonance that is reflected by every heartbeat, then we are in tune with the rhythm of the universe. It happened during the Civil Rights movement when the songs kindled the resistance; it happened in South Africa when the freedom songs and the toyi toyi dances toppled apartheid.

Wheatley OK, I talked about shifting realities when I became a Christian, but isn't it a shift in consciousness to see one's place in the world from a cosmological perspective? It changes everything to realize that we have specific tasks to accomplish and that we live in the midst of processive creations that continue to be the source of our artistic inspiration. Art is prophetic when it speaks to and depicts both the potential and the illness in society. We live in an era of terminal self-absorption. Even our approach to spiritual matters focuses on what we are going to get from the experience. It is a form of sickness to place oneself at the center of the universe and to expect that things will "work together for personal good."

Langston And the best antidote for that disease is to encounter the spiritual through the arts, to identify issues of humankind through drama, to lament through Billie Holiday's songs, and to find a place of joy that only the Step Brothers could get you to with their feet. But I'm not certain that a cosmological view is an option for those caught in webs of everyday earthbound existence. Laundry and grocery shopping and the business of life and death demand too much of our time to be easily dismissed. But if we extend ourselves beyond the boundaries of our individual lives, there just may be a moment of transcendence where the human spirit vaults beyond itself toward another dimension of understanding.

Brooks It is a leap that is difficult to take. Religious folks argue that transcendence is the business of the church, but once institutionalized, places of worship become centers of stasis. They can't offend; they must sustain themselves financially and satisfy their constituencies. But isn't that what art does? It takes you somewhere, even if you don't like the place that you end up. There is in art what Matthew Fox called "a fierce imagination" that sustains folks during times of suffering and that heals in unexpected ways.

> The artist . . . cleanses and recycles the toxins in a culture. Artists turn pain into insight and struggle into triumph. . . . Artists add awe to awe and beauty to beauty and wonder to wonder. . . . The artist is to the community or body politic what the liver is to the human body: a cleanser and recycler of waste and toxins.[13]

Tupac Art does what churches often cannot. It pushes the vision toward the future without concern about the status quo. I'll give you an example. Right after the Civil Rights movement, the songs were about becoming the beloved community. And yet, an entire generation was left behind while parents reached for the brass ring and wore themselves out working themselves into debt and the illusion of inclusion. There seemed to be no way out until the rhythm of the imagination of abandoned children and their new gang families burst onto the scene. We could have all pretended that everything was going to be great someday, but instead the rhythm of the drum in the beat of the tongue declared a new order.

Brooks I agree, Pac, but what makes me sad is that a particular art form was responsible for short-circuiting the lives of artists like you and Biggie. I don't know any watercolor artists or ballet dancers dying for their art. Was this supposed to be cool?

Tupac No, Sister, but it was destiny. The end was written into the beginning, from the first rap to the emergence of a gangsta theme for the genre. True, gangsta mentality gave our anger a shape, but it was a shape that directed us toward a violent end. We didn't start the violence; it was in the culture. The violence was in the American imagination. It had played out with every minority encountered. But what we did not know was that we were in the cross hairs and that the energy generated by the rhythm of our tongues was enough to start a revolution.

Think about it. If our anger had not been channeled into fantasy gangsta wars, we might have ignited the youth in ways that would not have served the interests of the political power brokers. We were exposing the invisible underbelly of the beast of American culture. The beast fed on a capitalistic rapaciousness that ground the poor to powder and authorized the police to maintain the status quo "by any means necessary." The isolation of poor people of color was complete.

Brooks I know that you're right about this, Brother, because Reagan could say with a straight face that nobody in America went to bed hungry, and Barbara Bush could tell Katrina victims that they were better off homeless, bereft, and living in a coliseum. Rappers exposed the nasty in American "nice." But still, I'm looking for the liberation in the art, the words, dance, or novel that releases my soul.

Langston We don't have to force a connection between art and cosmology; it is a natural nexus. The rhythm of my poetry "What happens to a dream deferred" asked public questions about commonly held concerns.

Brooks And all of this time I thought you were talking raisins and such.

Langston You know better, Gwen. I was talking about politics as usual inside and outside of the community. I was talking about our limited vision and the need to plan beyond freedom celebrations.

Writer So what is the wisdom that you leave the next generation?

Langston That in the midst of social crisis, there will be a rhythm of life that connects us to something greater. It may be in the hum of your grandmother's prayer or the beat of a poem that dares to imagine a different order. Art invites you to laugh when lament makes more sense, to write poetry when target practice seems more prudent. Art draws the boundaries of the life space with broad, erasable strokes; it encompasses laughter and never takes itself too seriously. It reflects back to you the person that you hardly even recognize in the mirror but that God recognizes as the created and creating spirit.

Tupac I want my homies to know that the rhetoric of conformity never saved anyone from the brutal effects of self-delusion. The universe takes voice in our throats, but the sound is not of our making. We can enjoy and learn from it, but to claim that it originates out of our limited embodiment is absurd. I would urge young people to prophesy to the beat of their own rhythms, to speak using the language at hand.

I guess that I have to qualify that advice. In the twenty-first century, everyone is speaking about everything. First, find your voice. It may not match the sound of any other voice, but if it is yours it will be unique and you will know that you are amplifying deeper connections.

Wheatley No matter how confusing, there is an inner map, an intuition or inkling. Follow it!

Brooks And I would urge this next generation to remember us so that they can build on the foundation that we began, and so that they don't have to repeat our mistakes. I would also urge them to connect to the larger community of humankind, beyond local neighborhoods, beyond global villages, toward a meaning-making life space within a vast and generative universe.

Langston And while you are struggling for justice and righting wrongs, don't forget that the moral center includes spaces for dance and song and a poem or two. Art will remind you when you are working that second and third shift that you are not a "worker" cog in the global marketplace. The best thing about art is that it will remind you that you are free to create your future. A dance, if only a wriggle at the water cooler, will ignite sense memory regarding your connections to a world in motion, a witty saying at the bottom of a birthday card will put you in league with prize-winning writers, and a song will pluck the waiting cords of a vibrating and resonant cosmos.

Tupac And if, like Vuyisile Mini, songwriter and activist hanged in South Africa during the resistance to apartheid, you cannot escape your fate, then sing on the way to the gallows.

Brooks Harmonize on the way to Gethsemane.

Langston Let art embrace you until freedom comes.

Summary: What Is the Role of Prophetic Art and Artists in the Quest for Liberation?
Writer We are looking for a liberator to lead the way, and Pac, Langston, Gwendolyn, and Phillis are suggesting that art will suffice. So what do we do? Do we wait for the next blast of a tenor sax or sidle up to the fire of molten glass slumping toward shape? The world is in such bad shape that I wonder whether a brush stroke or a concise poem will suffice.

Sarah All the more reason to offer art as a counterintuitive response to violence and injustice; nothing else seems to have worked. The artists of the twenty-first century have the responsibility to break through the silences and to reject the prevailing lies and political promises. Instead, they must photograph war-torn lands and paint the truth about governmental neglect during the hurricane devastation in New Orleans. They must put the tools of liberation in the hands of youth. It's difficult to fire a gun at a rival gang member when you are looking through the lens of a digital camera. Artists can't take anyone's word that freedom has come; they must test the assertion with oil paint, fiery sermons, and the dialogue of a play.

Writer Creativity reflects the desire and imagination of the universe on behalf of the human community. For certain, some of Coltrane's notes exist nowhere except in far-away galaxies. Rap reflects the beat of discontent in the post–civil rights generation and the drum denied during slavery, but it is also the pulsing of light and energy transformed.

The Exercises

1. What does it mean to be a prophetic artist? Write a five-point guide that describes what the arts can accomplish in the quest for liberation.

2. Can you name prophetic artists who are alive today? How do they differ or connect to the conversation in this chapter?

3. Choose the lyrics from one of Tupac's songs and compare it to one of Gwendolyn's or Langston's poems. Do they imagine liberation in the same ways?

4. Identify an artistic gift in your own life. If you believe that you don't have one, choose an artistic discipline that you would like to explore. Describe in one page how you would use this gift to "speak truth to power." Then create something. Let the transformative power of artistic expression flow freely through you.

5. Congregational art takes interesting forms. Describe the best and the worst of artistic expression in the church.

6. Choose a controversial issue for the church and describe how an artistic approach to that issue can improve the dialogue (for example, the HIV/AIDS quilt), or choose a creative project that addresses the issue.

For Further Reflection

Tupac Shakur

Alexander, Frank, and Heidi Siegmund Cuba. *Got Your Back: Protecting Tupac in the World of Gangsta Rap.* New York: St. Martin's Griffin, 2000.

Batsfield, Darren Keith. *Back in the Day: My Life and Times with Tupac Shakur.* New York: Ballantine, 2002.

Brooks, Darren. *Maximum 2Pac: The Unauthorised Biography of 2Pac.* New York: Chrome Dreams, 2003.

Chopmaster J. *Static: My Tupac Shakur Story.* San Francisco: Herb 'N Soul, 1999.

Constantine, Alex. *The Covert War against Rock: What You Don't Know about the Deaths of Jim Morrison, Tupac Shakur, Michael Hutchence, Brian Jones, Jimi Hendrix, Phil Ochs, Bob Marley, John Lennon, the Notorious B.I.G.* Venice, Calif.: Feral House, 2000.

Datcher, Michael, Kwame Alexander, and Mutulu Shakur. *Tough Love: The Life and Death of Tupac Shakur, Cultural Criticisms and Familial Observations.* Alexandria, Va.: Alexander, 1997.

Dyson, Michael Eric. *Holler If You Hear Me: Searching for Tupac Shakur.* New York: Basic Civitas, 2001.

Frokos, Helen. *Tupac Shakur (They Died Too Young).* Philadelphia: Chelsea House, 2000.

Gilmore, Mikal. "Tupac Shakur: Easy Target." In *Night Beat: A Shadow History of Rock and Roll: Collected Writings.* London: Picador, 1998.

Gobi. *Thoughts on Tupac Shakur in Pictures and Words.* New York: Simon & Schuster, 2003.

Hoye, Jacob, and Karolyn Ali. *Tupac Resurrection, 1971–1996.* New York: Pocket, 2003.

Jones, Quincy, and the Editors of Vibe Magazine. *Tupac Amaru Shakur, 1971–1996.* New York: Crown, 1997.

Scott, Cathy. *The Killing of Tupac Shakur.* Las Vegas, Nev.: New Huntington, 1997.

Shakur, Tupac A. *The Rose That Grew from Concrete.* New York: Simon & Schuster, 1999.

———. *The Tupac Shakur Collection*. New York: Warner Bros. Publications, 2001.
Sullivan, Randal. *Labyrinth: A Detective Investigates the Murders of Tupac Shakur and Biggie Smalls, the Implication of Death Row Records' Suge Knight, and the Origins of the Los Angeles Police Scandal*. New York: Atlantic Monthly, 2002.
White, Armond. *Rebel for the Hell of It: The Life of Tupac Shakur*. New York: Thunder's Mouth, 1997.
(Resources adapted from http://www.hiphoparchive.org/about/roundtables/all-eyez-on-me/tupac-selected-bibliography [accessed June 8, 2008])

Phillis Wheatley

Bennett, Paula. "Phillis Wheatley's Vocation and the Paradox of the 'Afric Muse.'" *PMLA* 113, no. 1 (1998): 64–76.
Bly, Antonio T. "Wheatley's 'on the Affray in King Street.'" *Explicator* 56, no. 4 (1998): 177–80.
———. "Wheatley's 'on the Death of a Young Lady of Five Years of Age.'" *Explicator* 58, no. 1 (1999): 10–13.
———. "Wheatley's 'to the University of Cambridge, in New-England.'" *Explicator* 55, no. 4 (1997): 205–8.
Burke, Helen. "Problematizing American Dissent: The Subject of Phillis Wheatley." In *Cohesion and Dissent in America*, ed. Carol Colatrella and Joseph Alkana, 193–209. SUNY Series in American Literature. Albany: State University of New York Press, 1994.
Cima, Gay Gibson. "Black and Unmarked: Phillis Wheatley, Mercy Otis Warren, and the Limits of Strategic Anonymity." *Theatre Journal* 52, no. 4 (2000): 465–95.
Derounian, Kathryn Zabell, and William H. Robinson, eds. *Critical Essays on Phillis Wheatley*. Boston: Hall, 1982.
Gates, Henry Louis, Jr. "From Wheatley to Douglass: The Politics of Displacement." In *Frederick Douglass: New Literary and Historical Essays*, ed. Eric J. Sundquist, 47–65. Cambridge Studies in American Literature and Culture. Cambridge: Cambridge University Press, 1991.
———. "Phillis Wheatley and the Nature of the Negro." In *Critical Essays on Phillis Wheatley*, ed. William H. Robinson, 215–33. Critical Essays on American Literature. Boston: Hall, 1982.
Grimsted, David. "Anglo-American Racism and Phillis Wheatley's 'Sable Veil,' 'Length'ned Chain,' and 'Knitted Heart.'" In *Women in the Age of the American Revolution*, ed. Ronald Hoffman and Peter J. Albert, 338–44. Charlottesville: University Press of Virginia, 1989.
Isani, Mukhtar Ali. "The British Reception of Wheatley's Poems on Various Subject." *Journal of Negro History* 66, no. 2 (1981): 144–49.

———. "The Methodist Connection: New Variants on Some Phillis Wheatley Poems." *Early American Literature* 22, no. 1 (1987): 108–13.

———. "Phillis Wheatley and the Elegiac Mode." In *Critical Essays on Phillis Wheatley*, ed. William H. Robinson, 208–14. Critical Essays on American Literature. Boston: Hall, 1982.

Jordan, June. "The Difficult Miracle of Black Poetry in America; Or, Something like a Sonnet for Phillis Wheatley." *Massachusetts Review* 27, no. 2 (1986): 252–62.

Levernier, James A. "Phyllis Wheatley and the New England Clergy." *Early American Literature* 26, no. 1 (1991): 21–38.

———. "Wheatley's 'On Being Brought from Africa to America.'" *Explicator* 40, no. 1 (1981): 25–26.

Mason, Julian D., Jr., ed. *The Poems of Phillis Wheatley*. Rev. and enl. ed. Chapel Hill: University of North Carolina Press, 1989.

O'Neale, Sondra. "A Slave's Subtle War: Phyllis Wheatley's Use of Biblical Myth and Symbol." *Early American Literature* 21, no. 2 (1986): 144–65. (especially recommended)

Poems on Various Subjects Religious and Moral by Phillis Wheatley, Negro Servant to Mr. John Wheatley of Boston, in New England, Printed for A. BELL, Bookseller, Aldgate; and sold by Messrs. C O X and B E R R Y, King-Street, *BOSTON*. MDCCLXXIII.

Robinson, William H. *Critical Essays on Phillis Wheatley*. Boston: Hall, 1982.

Rogal, Samuel J. "Phillis Wheatley's Methodist Connection." *Black American Literature Forum* 21, no. 1/2 (1987): 85–95.

Silvers, Anita. "Pure Historicism and the Heritage of Hero(in)es: Who Grows in Phillis Wheatley's Garden?" *Journal of Aesthetics and Art Criticism* 51, no. 3 (1993): 475–82.

Sistrunk, Albertha. "The Influence of Alexander Pope on the Writing Style of Phillis Wheatley." In *Critical Essays on Phillis Wheatley*, ed. William H. Robinson, 175–88. Critical Essays on American Literature. Boston: Hall, 1982.

Walker, Alice. "In Search of Our Mothers' Gardens: Honoring the Creativity of the Black Woman." *Jackson State Rev* 6, no. 1 (1974): 44–53.

Willard, Carla. "Wheatley's Turns of Praise: Heroic Entrapment and the Paradox of Revolution." *American Literature: A Journal of Literary History, Criticism, and Bibliography* 67, no. 2 (1995): 233–56.

Langston Hughes

Hughes, Langston. *The Best of Simple*. Illustrated by Bernhard Nast. New York: Hill and Wang, 1961.

———. *The Best of Simple*. Sound recording. Folkways Records, 1961.
———. *Laughing to Keep from Crying*. 1st ed. New York: Holt, 1952.
———. *Laughing to Keep from Crying and 25 Jesse Simple Stories*. Limited ed. Franklin Center, PA: Franklin Library, 1981.
———. *Not Without Laughter*. 1930. Reprint, New York: Collier, 1979.
———. *The Return of Simple*. Ed. Akiba Sullivan Harper. 1st ed. New York: Hill and Wang, 1994.
———. *Short Stories*. Ed. Akiba Sullivan Harper. 1st ed. New York: Hill and Wang, 1996.
———. *Simple Speaks His Mind*. 1950. Reprint, Mattituck, N.Y.: Aeonian, 1976.
———. *Simple Stakes a Claim*. New York: Rinehart, 1957.
———. *Something in Common and Other Stories*. New York: Hill and Wang, 1963.
———. *Tambourines to Glory: A Novel*. 1958. Reprint, New York: Hill and Wang, 1970.
———. *The Ways of White Folks*. 1st ed. New York: Knopf, 1934. Reprint, New York: A. A. Knopf, 1969; Vintage, 1971.

Gwendolyn Brooks
By Gwendolyn Brooks
Annie Allen (1949), poetry collection
Maud Martha (1953), novel
Bronzeville Boys and Girls (1956), illustrated collection of poems for children
The Bean Eaters (1960), poetry collection
Selected Poems (1963), poetry collection
In the Mecca (1968), poetry collection
Riot (1969), poetry collection
Family Pictures (1970)
Aloneness (1971)
Report from Part One, autobiography (1972)
The Tiger Wore White Gloves, children's book (1974)
Beckonings (1975)
Primer for Blacks (1980)
To Disembark (1981)
Young Poet's Primer (1981)
Mayor Harold Washington; and Chicago, the I Will City (1983)
The Near-Johannesburg Boy and Other Poems (1986)
Blacks (1987)
"Gottschalk and the Grand Tarantelle" (1988)
Winnie (1988)
Children Coming Home (1991)

By Others
Brown, Patricia L., Don L. Lee, and Francis Ward, eds. *To Gwen with Love: An Anthology Dedicated to Gwendolyn Brooks.* Chicago: Johnson, 1971.
Davis, Arthur P. *From the Dark Tower: Afro-American Writers (1900 to 1960).* Washington, D.C.: Howard University Press, 1974.
Kent, George E. *A Life of Gwendolyn Brooks.* Lexington: University Press of Kentucky, 1990.
Kufrin, Joan. "Gwendolyn Brooks." In *Uncommon Women*, 35–51. Piscataway, N.J.: New Century, 1981.
Loff, Jon N. "Gwendolyn Brooks: A Bibliography." *College Language Association Journal* 17 (September 1973): 21–32.
Madhubuti, Haki R., ed. *Say That the River Turns: The Impact of Gwendolyn Brooks.* Chicago: Third World, 1987.
Melhem, D. H. "Gwendolyn Brooks: Humanism and Heroism." In *Heroism in the New Black Poetry: Introductions and Interviews*, 11–38. Lexington: University Press of Kentucky, 1990.
———. *Gwendolyn Brooks: Poetry and the Heroic Voice.* Lexington: University Press of Kentucky, 1987.
Miller, R. Baxter. *Langston Hughes and Gwendolyn Brooks: A Reference Guide.* Boston: Hall, 1978.
Mootry, Maria K., and Gary Smith. *A Life Distilled: Gwendolyn Brooks, Her Poetry and Fiction.* Urbana: University Press of Illinois, 1987.
Shaw, Harry B. *Gwendolyn Brooks.* Boston: Twayne, 1980.
Wright, Stephen Caldwell. *The Chicago Collective: Poems for and Inspired by Gwendolyn Brooks.* Sanford, Fl.: Christopher-Burghardt, 1990.
———. *On Gwendolyn Brooks: Reliant Contemplation.* Ann Arbor: University of Michigan Press, 1996.
(Resources adapted from http://www.math.buffalo.edu/~sww/brooks/brooks-biobib.html#bio)

Tolerance, like any aspect of peace, is forever a work in progress, never completed, and if we're as intelligent as we like to think we are, never abandoned.
—Octavia Butler

9. Beyond Mountaintops: Toward a Long-Awaited Future

The human spirit is resilient and . . . that truth—no matter how long you abuse it and how long you try to crush it—will, as Dr. King would say, rise up again, and in the final analysis will prevail. From the point of view of the poor, the hungry, the disenfranchised, the wretched of the earth . . . there will never be peace until their condition has been alleviated and until their humanity is in full bloom.
—Harry Belafonte

In this chapter, three giants of the twentieth century talk about the future of African Diasporan people in the United States. It begins with a conversation with W. E. B. DuBois, Martin Luther King Jr., and Mary McLeod Bethune. Fannie Lou Hamer and Shirley Chisholm make a brief appearance. The topics include globalization and the responsibility of African Diasporan people to consider the ongoing struggle for justice as an intercultural and cosmological enterprise.

The Elders
Martin Luther King Jr., 1929–1968

We caused the sagging walls of segregation to come down, and yet, in spite of a decade of significant progress, the problem is far from solved. The deep rumbling of discontent in our cities is indicative of the fact that the plant of freedom has grown only a bud and not a flower.[1]

Love is the only creative, redemptive, transforming power in the universe.[2]

Martin Luther King Jr. did not grow up planning to have a national holiday or a sound bite on his birth and death days. He followed the path of his life as most of us do. He came from an influential line of preachers and followed that trajectory. He earned his Ph.D. from Boston University; married the love of his life, Coretta Scott; and accepted a pulpit appointment at Dexter Avenue Baptist Church in Montgomery, Alabama, of all places. In accordance with the wisdom of the universe, he was in the right place at the right time when the Civil Rights movement ignited. The climate was ripe for social change. Young and old were weary from living in a dangerously segregated society. The marches followed, and so did the arrests and threats on King's life.

King believed that the universe had a moral arc that bent toward justice. His responsibility was to prophesy to a nation comfortable with injustice that a new day had come. At age thirty-five, he received the Nobel Peace Prize. King's message of love as a modality for social change was not limited to issues of race; he also focused on war and economic injustice. On April 4, 1968, he was in Memphis, Tennessee, to support the garbage workers' strike and was assassinated on the balcony of the Lorraine Motel. He preached persistence and hope to the downtrodden, spoke reconciliation and unwavering resolve to the dominant culture, and died with prophetic insight regarding the dawning of a long-awaited future.

Mary McLeod Bethune, 1875–1955

I leave you love. I leave you hope.
I leave you the challenge of
developing confidence in one another.
I leave you a thirst for education.
I leave you respect for the uses of power.
I leave you faith. I leave you racial dignity.
I leave you a desire to live harmoniously with your fellow men. [sic].
I leave you finally a responsibility to our young people.[3]

In an era when children are dying and indifference rules, I wonder how many people have read this legacy. Bethune was born one of seventeen children in Mayesville, South Carolina. She rose to prominence as an educator and activist for women's rights. Bethune was founder of the Daytona Normal and Industrial Institute for Negro Girls (now Bethune-Cookman College). She served as president of this institution for thirty-nine years. She cared about the education of young people because her own life changed when she was given the opportunity to study in Mission schools and at the Moody Bible Institute. After she was educated, she moved to Daytona Beach to start her own school. The school was a one-room site of opportunity for young women.

Bethune was also a leader in the black club women's movement, a founding member of the National Council of Negro Women, and a dedicated public servant and advisor to presidents. Her appointments include director of Negro affairs in the National Youth Administration from 1936 to 1944, consultant to the U.S. secretary of war, and consultant on issues of race and ethnicity at the charter conference of the United Nations. She was a woman of stature and grace who envisioned liberation as a vital outcome of education.

William Edward Burghardt (W. E. B.) DuBois, 1868–1963

In a peculiar way, then, the Negro in the United States has emancipated democracy, reconstructed the edifice of freedom and been a sort of eternal test of the sincerity of our democratic ideals.[4]

W. E. B. DuBois was born in Barrington, Massachusetts. It might seem that being born in a northern state devoid of the more virulent forms of racism would dull his fervor for freedom, but this was not the case. His focus on justice in America was intense and persistent. He was inclined toward intellectual pursuits and attended Fisk and Harvard. He pursued his doctoral degree in Europe (at the University of Berlin, Germany) until his financial support was withdrawn. He returned to Harvard to complete the degree. DuBois taught at Wilberforce University, the University of Pennsylvania, and Atlanta University.

He is known for his scholarly contributions to the social sciences, his research on the social conditions of Negro communities, and his Pan-Africanism and ideological struggles with Booker T. Washington and Marcus Garvey. Although he was a founder of the Niagara Movement (an organization formed to address issues of discrimination in the black community), the group could not sustain itself due to organizational ideological rivalries. DuBois was also a founding member of the NAACP, but he eventually moved toward a Pan-Africanist view that did not include integration as a goal. Ultimately, he left the United States to live his final years in Accra, Ghana. He died on the eve of the March on Washington in 1963.

Shirley Chisholm, 1924–2005

An increasing number of black women are beginning to feel that it is important first to become free as women, in order to contribute more fully to the task of black liberation.[5]

Shirley Chisholm was born in New York to parents of Caribbean descent. We know her primarily for her political tenacity and her willingness to fight a losing battle to become the first woman candidate for the presidency of the United States. She was a congresswoman from the 12th District of New York and was a founding member of the black caucus as well as a mediator of cultural issues across the racial divide. When she ran for the presidency of the United States, she knew that she would have to contend with the politics of race and gender, but she ran anyway. Although she wanted to be seen as a viable candidate, she was willing to embody liberation in the imaginative construction of a nation still struggling with its discomfort with difference.

The Conversation

As they say in my part of the country, I was in deep cotton. The elders assembling for this conversation framed my own understanding of leadership. I grew up watching them engage God, community, and culture. They dreamed impossible dreams and then woke up and told everybody that "trouble don't last always." They led us on an uncharted path toward freedom with wisdom, humility, and prophetic action. We saw it with our own eyes, sitting in front of black-and-white televisions transfixed by their stalwart commitments, and we knew that the work of liberation was difficult but doable.

DuBois When are you all going to get wise and get out?

MLK You can't break the back of oppression unless you stay.

Hamer Talk about staying, y'all, I was in the "body" chapter [chapter 6] and can't stay to talk with you, but I just wanted to say that I put my back on the line. I don't know about the back of oppression, but mine sure took a whuppin'.

MLK I never saw a woman who could handle herself like you did, Fannie.

Hamer Maybe that's because you weren't looking, Martin. I wasn't one of those prissy women, but I wasn't one of a kind. There were women like me having babies in fields and still picking their share, saving their families, and keeping the faith. They put their backs into the support of their communities and sacrificed their own ideas about liberation for the cause of unity.

DuBois That's all well and good, but it is not the beaten back that saves a nation; it's the intellectual capital that advances the race.

Hamer You don't say! I'm glad that Tubman didn't hear that. She freed people by getting up and walking. I've got nothing against higher learning, but sometimes you have to put your hand to something. You have to tear down a wall or raise a child or bake some biscuits, something real and tangible to stave off the sadness. You don't overcome if you don't keep moving.

MLK You really can't criticize Fannie, W. E. B. You left; she stayed.

DuBois Don't start something that you can't finish, Martin. I never ran from anything in my life, and you know it. You're just trying to get me riled so that I'll tell my Africa stories.

MLK You know me, don't you?

DuBois I left because I wanted to spend a few truly free days in the motherland and I wanted to die there.

MLK We all want to find our way home, but the motherland is not a geographical location. You carry it within you. When I said that I'd been to the mountaintop, I was using an allusion to make a point but, on another level, the mountain rose up to meet me. During that final speech, I was climbing and sweating and talking with God. One thing that I came to realize during my life sojourn was that our limited access to the available dimensions could be easily transcended through prayer and invocation.

I understand that human beings currently have access to four dimensions, height, length, and width plus time, yet as of this writing there are eleven or more available. Although scientists say that the others can't be accessed, I watched my mother and the church mothers do it all of the time. The prophets crossed back and forth at will. The boundaries between the spiritual and scientific layers of reality are permeable.

Hamer I don't know much about the science of multiple dimensions, but I was able to "escape" the beatings and abuse by leaving the site of the abuse. I would travel in my mind to another place, but I never thought of that space as dimensional. It was a place where their blows could not reach my spirit. There, I could have a little talk with Jesus and tell him all about my troubles.

DuBois It sounds like she's going to break out in song.

MLK That'd be all right with me. Sing on, sweet sister, sing on . . .

DuBois Fannie, while you were talking to Jesus, did you get a clue as to why black folks were reaping the wrath of heaven and hell?

Fannie Lou Now what would I look like demanding answers about my little ol' beating from a crucified Savior? As a people, we suffer, but so did many people before us, and I don't see any end in sight. Anyway, we are supposed to be talking about the liberation project and how it turned out.

MLK Actually, we're right on target. The subject is liberation and the cosmos. I don't know much about the stars, but I do know what it means to be free.

DuBois Before we resume that discussion, I'd like to get some clarity about the goals of the liberation movement. Integration puzzled me. I was never sure why you wanted to fit into white society.

MLK It was about getting access to what we'd earned by the sweat of our backs and to a future that was open to all possibilities. As far as integration goes, I'm not so certain that I understand what that word means anymore. It is a word that has not served us well over the years. Inclusion may offer a better description of our desire for a recognized place in the human family. But it's a legitimate question. At the time, we saw integration as the opportunity to participate in Western society. We didn't expect paradise, but we did expect to have our humanity respected and recognized.

DuBois The changes in society happened so quickly that it was almost impossible to factor the meaning of those changes into our personal and communal goals. One moment, all of the forces of the dominant systems are bearing down on the community to keep them "in their place," and then suddenly it's all over. The change that took place was legal and apparent, but the changes that were needed took much longer than the desegregation of lunch counters. Without acknowledgment of the past, it was as if we were waking from a dream. Did we dream slavery? By agreeing not to discuss such matters, we invited revisionist history. A truth and reconciliation process was needed so that the stories could be told and remembered.

MLK While I am disappointed that this generation knows so little about those who went before, there have been surprising gains.

DuBois I don't agree. While things may be better, life in postmodern America is fragmented and confusing.

MLK But isn't public life in any context complicated? I hope that you didn't expect a perfect national context. W. E. B., I'm betting that when you arrived in Ghana you thought that you had entered the promised land, and that's because you were finally far enough away from American versions of oppression. But I'm sure that you learned that your new neighbors had oppressions of their own. Perhaps you just had to live there awhile to recognize the problems. But unless I miss my guess, what looked like paradise to you was just the absence of familiar domination systems.

DuBois I know that you're right, but it still felt good to leave and it felt even better to make my transition on eve of that great day when you claimed your stake in America on the mall during the March on Washington.

MLK You were on my mind when I gave the call for unity that was central to the movement.
   As a people, we are still on the journey toward the wholeness that the universe offers. If you ask me, the best way to access the mysteries of the universe is through the mysteries of faith. The stories of the Bible girded us for the early battles for justice and freedom because they were mystical and improbable.

> Religion, in this sense, is not simply a doctrine of faith, or the methods and practices of church, rather it is all the ways we remind ourselves of who we really are, in spite of who the temporal powers may say we are. Religion is how we situate ourselves, how we understand ourselves, in a particular place and time vis-à-vis ultimate reality, vis-à-vis God.[6]

Fannie Lou I'd like to stay and talk about ultimate realities, but I've got to go. Let me move aside and let another powerful spirit enter.

Bethune Brothers, how are you?

MLK Mother Bethune, what an honor. We were just talking about the future. I bet you have something to say about education and the state of the liberation project.

Bethune I do, but I'm not alone. Shirley is here.

DuBois Shirley Chisholm, Madam President . . . how are you? To what do we owe the pleasure?

Shirley C. What an opportunity! I just wanted to share a few thoughts, and then I have to leave. I will defer the rest of the conversation to Mother Bethune. I want future generations to know that my campaign for president of the United States was unsuccessful in terms of winning the national election, but on another level it gave the country a vision of what power would look like when the black women who raised them and cleaned their houses assumed political leadership.

I understand that there are many African American candidates for national political offices now, even for the presidency of the United States. This is my challenge to those potential leaders and to the next generation: do you have a model of leadership based on "vision" that differs in any way from politics as usual? The political handlers know only one way to hone a candidate, and that way requires packaging as well as dodging real conversation with real constituents. If there is to be diversity in public office, it should offer the nation an opportunity to benefit from a fresh viewpoint. Does your social location give you fresh insights about leadership and significantly different approaches to problem solving, or will you mimic what you have seen?

DuBois This is lofty talk about leadership, but what do you do when a racial/ethnic candidate is assailed because of the prophetic and liberationist rhetoric of his or her pastor?

Shirley C. You respond in the same way that you must when a woman candidate is called "difficult" because she talks tough and brings her best game to a political race. Whoever you are, it's still about vision. I'm waiting for a political leader to arise who exemplifies a fearless, peaceful, and visionary leadership. Don't blame the people for losing interest in the campaigns of candidates who are all wearing the same mask and muttering the same phrases. The people need choices. With regard to religious participation in national politics, I believe that the body politic includes all of us. Religious leaders who decide to exercise prophetic gifts in the service of an improved political landscape have the right to voice their opinion. When you have diminished public discourse, any voice that challenges the status quo will be deemed dangerous and out of order. But that is the subject of another conversation. I have to leave. Blessings to all of you. You are certainly living during interesting times.

Bethune I'm in this chapter to represent women, because they also want to be heard. During the Civil Rights movement, they stayed in the background, in the belief that it was more important that the community move forward. They reckoned that there would be time to assert their own views of liberation. That time has finally come, and they want me to represent their interests during this powerful conversation. They want to suggest to another generation that the only way forward is to stop the charade, the pretense of wellness and success. They want me to tell you that lament is needed as a ritual of cleansing and preparation for what is yet to come. It is a step in the process of liberation that was never completed.

DuBois Oh, so now we have time for a good cry? If this is all that the women have to offer, I suggest that they regroup and come up with something better.

Bethune Don't go there, Brother. The generations that have followed slavery have been crying throughout their lifetimes; they have just chosen to do it on the inside. Their spirits are riddled with the salt of unreleased tears.[7]

> To lament is not simply to grieve or mourn. Biblical lament, faith's outcry to God in the grip of trouble, is a rhetoric that wails and rages, protests and interrogates, and finally whispers its hope.[8]

MLK The whispered hope that echoes through every wail and cry of anguish is that the troubles of this world are not the end of the story. Now we see through a glass darkly and not face to face. This generation is inundated with twenty-four-hour news stations that bring the pain of the world into your living rooms. Yet, your lives in Western societies seem to go on unchanged. You are inundated with news of disaster and death, yet even in your compassion, you seem distanced and detached from the grit and horror going on in the world. Some of you respond in the best ways that you can, but within weeks, even the most astonishing events are grist for late-night talk show jokes.

Bethune We know the healing power of laughter, but do we know when and how to weep? During Katrina, shocking television images held the nation suspended in disbelief for a few days, but then everything returned to "business as usual," as talking heads on both sides of the political spectrum drove the sounds and sights of dying people from the public consciousness. Survivors were shipped to waiting relief in other cities, but because the bodies were not counted for the world to see, there was nothing to weep over.

MLK In every generation, there seems to be one war or another, but lately you hear about the deaths but are not allowed to see the coffins. It's hard to lament in the abstract, and without lament there can be no joy. Lament carves pathways toward a different vision of life in community. It vocally acknowledges that things are not as they should be. Lament deconstructs the tents of superficial contentment by allowing the hidden sense of abandonment to emerge. In the stillness that follows the wail comes the opportunity for amazement, letting go and finally healing and resisting through an activism that prioritizes compassion and justice.[9]

Bethune That is why my call to the next generation is to reclaim the possibility of real joy through the healing practice of lament. I am suggesting that we weep with those who weep, that we moan over the harm done on our behalf and by our hands.

DuBois Even if I were to agree that tears are a necessary next step on the pathway to liberation, I'm not sure that a few sniffles would do it. Instead, if lament begins, it will unleash torrents of regret and sympathy, not just for your own concerns but for the concerns of a broken world. However, despite my misgivings, lament may be the only remaining antidote to the bitter mix of arrogance and self-congratulations that the West has adopted.

MLK I know that this is true, because even during the tragedy of September 11, the call was for shopping to jumpstart a flagging economy. Televised rituals of mourning barely tapped into the communal grief. Within days, the task of mourning was turned over to local congregations. It is not that local congregations shouldn't be the line of first defense in such catastrophic times; it is that we have spent so much time avoiding death and grief that even the church is ill-equipped for times such as these.

Imagine that after the towers came down on 9/11, the whole nation fell to its knees weeping for its losses and its enemies, repenting for its failures and imploring the God of heaven to be our fortress and shield. Imagine that, instead of protesting the war on terror, folks assembled in Washington, D.C., and silently wept on the mall. There will always be enemies. The issue is not the reality of evil or its pervasiveness but the weapons that we choose to defend and resist. Tears are the most powerful declarations of faith even in the midst of siege.

Bethune Please don't misunderstand. There is a season for everything. We are urging the next generation to allow lament to act as a release valve for pent-up rage and generational frustration, to use lament as a teaching device for the children, and to allow the time of comfort that follows lament to knit the community together despite its differences.

To be perfectly honest, in a culture of strong men and stronger women, everybody wants to be hard and tough. They teach their sons not to cry and discourage their daughters from displaying emotions. This generation rejects tears because they signal vulnerability. As the last superpower on the earth during this era, they have rejected vulnerability as a sign of peace to the world, as a liturgical response to violence, and as a Christlike example in the midst of a landscape of fear.

DuBois I understand the price that we pay when we don't acknowledge our own pain, but every action in the history of the African Diaspora had a double edge. We sang gospel songs, but they encoded messages about the routes north to safety. Where is the edge in lament?

Bethune Well, for one thing, it's not easy to sleep when people are wailing. We need the sounds of gnashing of teeth, of moans and distress, to wake us from our cultural slumbers. When we are fully awake, we realize that there is a spiritual dimension to life that we rarely acknowledge. The world is not just DVDs and stuff. The world is a place where the opportunity for great good and astonishing evil is always within reach.

MLK The sense that your generation is struggling with unseen forces is not an illusion; it is real. But there isn't anything you can do about it while you are asleep. Please understand that this sleep is more like hypnosis than anything else. This generation is encouraged not to deal with reality, encouraged to live for their own needs. Christ calls us to be good stewards of our resources; society tells us that we need to pimp our rides. Christ calls us to love our neighbors, which includes the poor. Society pretends that the poor don't exist or that poverty is a failure of character or work ethic. Faith tells us that our lives are about service, self-sacrifice, submission, and salvation. Society tells us that life is about ego, expenditures, excess, and extremes, needing more and more thrills to feel alive. Whose report will you believe? Will you even hear this report if you are asleep at the switch? [At this point, both Bethune and DuBois offered hearty "Amens."]

Bethune I promise you that tears can be revolutionary. As an educator, I know that when a society hides its tears, its children will suffer. We see this in the increase of teen suicide and depression. Our children become sad and alienated in the midst of a society that pretends that there is no reason to lament. Under such circumstances, grief can rupture both spirit and body.

MLK I understand, you are not asking for the cry of a single voice; it is the lament of the community that leads to healing. It may seem that you are few in number, that you don't have the strength or means to overcome systems of oppression and death. All you have are prayers, faith, and courage. Yet, with this alone and the God who never leaves us alone, you must act. You must act even when the story doesn't end well, and the stories of the twenty-first century have not been ending well.

Writer I have been quiet for awhile, but now I have to say that I'm a bit concerned that the women who hear your story will resent having lament laid at their feet. The women who have appeared at this called meeting have not been a particularly weepy lot. In fact, there were so many periods of history when their strength of purpose and ingenuity protected the community. I'm having a hard time envisioning Ida B. or Tubman in tears.

Bethune The call for lament is not an invitation to moping or sadness. It is a call for ritual reorientation. With or without tears, lament is a communal act of cosmological engagement. Ancestors on the continent of Africa knew this. Ritual celebration and lament helped to orient life cycles and mediate powerful emotions and crises. Now, here is a connection to the cosmos that is natural and therapeutic. There are no apologies forthcoming for slavery; we are conflicted as to how much of the consumer lifestyle to adopt; our children seem to be beyond our reach. This is enough of a crisis to call for action. Instead of the "I'm all right, you're all right" mentality, it is an expression of unity and strength to lament.

DuBois I can see where the unresolved grief that African Diasporan people have carried for generations impedes their ability to move on. Most don't ever recognize or acknowledge the deep places of sorrow that shadow them. As a people, we need to shift our focus. Facing the pain is a first step. Nothing else has worked. We were told that if we just worked hard enough, got enough academic degrees, and lived well, that those activities would cancel out the suffering of this and past generations. It doesn't work that way.

MLK Yes, I agree that we need to turn a corner as a community, and I'm hopeful that the women will lead the way. But what happens after the ritual of lament is complete?

Bethune Then the stories will flow again. We have our own stories that are not being told. Instead, we have adopted the televised "stories" that inspire competition, easy violence, and skewed ideas about beauty, love, and our responsibilities for one another. Whether you know it or not, there are griots among you in this new generation. They are poets, drummers, preachers, and singers. They are found in every walk of life, and they are waiting to write and share the stories that defy the conspiracy of silence that pervades this present age.

DuBois But how does this activity advance the cause of liberation and translate to the rest of the world? We have been so myopic in our quest for freedom. Maybe tears will bless and refresh the African Diaspora in the West, but what about the rest of the world? Have you seen the townships in South Africa, or the bomb-blasted neighborhoods of Iraq, the HIV/AIDS-devastated continent of Africa, and the abandoned children in Brazil and the Philippines? For me, a cosmological view of liberation has to include not just the story of one oppression against one community; it is the collective response to oppression everywhere. It is a decision as to how we will connect to global neighbors.

Bethune You are right, W. E. B. The models of leadership must be more expansive. But that's another subject. We've covered the need for restorative rituals. Now we need to teach and prepare the type of leaders who are willing to go beyond the purported safety of "systems." We are still training our young people to work within systems that prefer uniformity.

MLK Yes, even pastors are preparing to fit within one denominational system or another, fearful that their unique callings will exclude them from the net of care that the systems provide. I don't know what would have happened to the Civil Rights movement if I had copied familiar models of ministry. God seeks leaders who will understand the uniqueness of their call and obey.

From elementary school through college, we are training people to acquiesce. We teach them enough for them to know that they don't know anything. We also teach them to respect only the knowledge that they can get from books. The strength of their inner wisdom is so diminished by the time that they go out into the world that they are properly confused and cowed. W. E. B., you asked about our connections to global neighbors. I can tell you that our effectiveness as committed members of wider communities does not occur because we know how to connect.

Writer Are we talking cosmology yet?

MLK Yes, because cosmology is not just about stargazing; it is about a cognizance of contexts beyond our reach. My transition took place on the day that Apollo 6 launched. The significance of that event is that as I was moving to the realm of the elders, a new day was dawning and the universe and its mysteries were becoming a part of our shared public consciousness.

> What we know and will know of reality fails to exhaust it. . . .
> This inexhaustible depth of nature, its unfathomableness, opens us to the Divine's mystery.[10]

Our joy is tied to our freedom, and our freedom is seeded in a God-given liberation that transcends socially constructed ideas about race, gender, sexuality, or class. Once you see yourselves as children of the universe, all else is a matter of alignment. In this cosmological view, one need not exhaust oneself in ongoing scrimmages with discrimination and oppression. These are false constructs, paper tigers in a living system that we call earth.

Bethune Our task is to align body, mind, and soul with the God of the universe. All else will follow. The activist and prolific cultural critic bell hooks offers this:

> My hope emerges from those places of struggle where I witness individuals positively transforming their lives and the world around them. Educating is always a vocation rooted in hopefulness. As teachers we believe that learning is possible, that nothing can keep an open mind from seeking after knowledge and finding a way to know.[11]

What keeps us from being the best that we can be? What limits, false or real, hinder our sense of fulfillment? Only the prisons in our minds can accomplish that task. For those who are marginalized, the first act of resistance is the reconfiguration of their own self-esteem. "People who are conscious of their connections to the cosmos will not be deterred from full exploration of their gifts, because true liberation includes the ability to conceptualize freedom beyond social configurations."[12]

DuBois I'm really getting a solid understanding of the cosmological aspects of liberation. The restoration of a social order gone awry begins with the vision of self and community as embedded and vital to the continuance of a universe that is "intelligent and intelligible."[13] Restoration comes when we remember that we are articulated star dust, bearing in our bodies the story of the universe and engaging the mysteries of the Creator.

Bethune A few years back, when Bill Cosby was asked about how he would keep teens from killing other teens for their sneakers, he said, "Change the object of their desires." He was calling for a panoramic vision of the world that necessarily takes a person beyond the neighborhood, beyond familiarity, and beyond the shoes on one's own feet.

Writer I don't know that I agree with Cosby's overall perspective, but I understand what you mean. Do you have any last words for the next generation?

MLK You mean we didn't say enough while we were with you? There is a smallness about our lives that does not befit our origins or our context. When we become citizens of the cosmos, small oppressions, no matter how persistent, are like fleas on the back of a camel. Annoying to be certain, but not the determining factor as to the direction that the camel takes. I urge a renewed spiritual discernment as to the next directions. Freedom may not ring from every hill; instead, its manifestation may be as gentle as the flowering of another peace-loving generation committed to reflecting that orientation in the governments that they choose and the lifestyles that they adopt.

DuBois My last words are the same as the first. The next generations must become citizens of the world and cognizant of the cosmos. They cannot point to the failings of the dominant culture if they spend their energy battling for city blocks in one neighborhood. The disconnection from history, the universe and its realities, and the lives of others is a social distortion that threatens the future. If the story of the cosmos is one of quantum entanglement and connection, then we are responsible for the well-being of neighbors far and near.

Bethune Education will cure myopia. The more that we learn about who we are, the better options there will be for living into the future.

MLK I hope that the conversation will inspire the next generation. There have been too many martyrs along the way. It is time to live to tell the story . . . to survive . . . to raise children and prepare the next generation for the challenges that they will face.

Summary: What Will Liberation Look Like in the Future?

Writer I am overwhelmed by the wisdom and by the conversation of the elders who have come through. I guess my task is to find a way to translate the message to the communities still being sustained by fragments of hope.

Sarah What comes through in this chapter is the power of love that transcends the life/death continuum, and a wake-up call to connect beyond the "hood" and the nation.

Writer The post–civil rights generation lives in a closed loop of work and canned entertainment. Perhaps a cosmology of openness would allow space for dialogue and provide an antidote to the limited templates that mark everyday life.

Sarah Ashé.

The Exercises

1. Examine with care Dr. Mary McLeod Bethune's Last Will and Testament (see a copy on the bottom of page 168). Identify three important elements of her legacy, and relate them to current social situations. Or, write your own legacy. What do you wish to leave in terms of your hope for future generations?

2. Research and write a three-page summary of the work of the National Council of Negro Women.

3. Does Shirley Chisholm's or Barack Obama's presidential candidacy impact, negate, or change W. E. B. DuBois's description of the "color line" in *The Souls of Black Folks*?

4. What is the role of education in the pursuit of liberation? Choose one of the following educators to discuss the role of teaching for liberation: Parker Palmer (*The Courage to Teach*); Paulo Freire (*Pedagogy of the Oppressed* or *Pedagogy of Freedom*), or bell hooks (*Teaching to Transgress*).

5. Define Pan-Africanism and name advocates for the position who are alive today. How does Pan-Africanism advance or impede liberation movements?

6. Martin Luther King Jr. is considered by some to have been a prophetic preacher. Can you name preachers who have assumed a public and prophetic role as they address issues of race, class, and gender? How does their discourse advance or impede liberation movements?

For Further Reflection

Martin Luther King Jr.

Branch, Taylor. *Parting the Waters: America in the King Years, 1954–63*. New York: Simon & Schuster, 1989.

———. *Pillar of Fire: America in the King Years, 1963–65*. New York: Simon & Schuster, 1998.

Cone, James H. *Martin & Malcolm & America: A Dream or a Nightmare*. Maryknoll, N.Y.: Orbis, 1992.

Dyson, Michael Eric. *I May Not Get There with You: The True Martin Luther King, Jr.* New York: Free Press, 2000.

King, Martin Luther, Jr. *The Autobiography of Dr. Martin Luther King, Jr.* Ed. Clayborne Carson. New York: Warner, 2001

———. *A Knock at Midnight: Inspiration of the Great Sermons by Rev. Dr. Martin Luther King, Jr.* New York: Intellectual Properties Management (Warner Brothers), 1998.

———. *The Papers of Martin Luther King, Jr.* Vol. 4, *Birth of a New Age, December 1955-December 1956*, ed. Clayborne Carlson. Berkeley: University of California Press, 1992.

———. *Strength to Love.* Philadelphia: Fortress Press, 1981.

———. *Stride toward Freedom.* New York: Harper & Row, 1958.

———. *Trumpet of Conscience.* New York: Harper & Row, 1968.

Mary McLeod Bethune

Bennett, Carolyn LaDelle. *Annotated Bibliography of Mary McLeod Bethune's "Chicago Defender" Columns, 1948–1955.* Lewiston, N.Y.: Edwin Mellen, 2001.

Hicks, Florence, ed. *Mary McLeod Bethune: Her Own Words of Inspiration.* Washington, D.C.: DARE Books, 1975.

Holt, Rackham. *Mary McLeod Bethune: A Biography.* Garden City, N.Y.: Doubleday, 1964.

Mary McLeod Bethune Papers: The Bethune Foundation Collection, Bethune-Cookman College, Daytona Beach, Florida.

Papers 1932–1942 (505 items), Amistad Research Center, Tulane University, New Orleans, Louisiana.

Records 1935–1983 for National Council of Negro Women, National Archives for Black Women's History, Mary McLeod Bethune Council House National Historic Site, National Park Service, Washington, D.C.

Sterne, Emma Gelders. *Mary McLeod Bethune.* New York: Knopf, 1957.

W. E. B. DuBois

Byerman, Keith. *Seizing the Word: History, Art, and Self in the Works of W. E. B. DuBois.* Athens: University of Georgia Press, 1994.

De Marco, Joseph. *The Social Thought of W.E.B. DuBois.* Lanham, Md.: University Press of America, 1983.

Duberman, Martin. "Du Bois as Prophet." *New Republic*, March 23, 1968, 36–39.

DuBois, W. E. B. *The Autobiography of W.E.B. DuBois: A Soliloquy on Viewing My Life from the Last Decade of Its First Century.* New York: International, 1968.

———. *Darkwater: Voices from within the Veil.* 1920. Reprint, New York: Schocken, 1969.

———. *Dusk of Dawn: An Essay toward an Autobiography of a Race Concept.* 1940. Reprint, New York: Schocken, 1968.

———. *The Fight for Equality and the American Century, 1919–1963.* New York: Henry Holt, 2000.

———. *The Souls of Black Folk*. Chicago: A.C. McClurg, 1903. Reprinted. in *W. E. B. DuBois: Writings*. New York: Viking/Library of America, 1986.

Dudley, David L. *My Father's Shadow: Intergenerational Conflict in African-American Men's Autobiography*. Philadelphia: University of Pennsylvania Press, 1991.

Horne, Gerald. *Black and Red: W. E. B. DuBois and the Afro-American Response to the Cold War, 1944–1963*. Albany: State University of New York Press, 1986.

Howe, Irving. "Remarkable Man, Ambiguous Legacy." *Harper's*, March 1968, 143–49.

Jones, Lombard C. "Fifty Years of Crusading for the Negro in America." *New York Times Book Review*, December 22, 1940, 16.

Lewis, David Levering. *W. E. B. DuBois: Biography of a Race 1868–1919*. New York: Henry Holt, 1993.

Manning, Marable. *W. E. B. DuBois: Black Radical Democrat*. Boston: Twayne, 1986.

Rampersad, Arnold. *The Art and Imagination of W. E. B. DuBois*. New York: Schocken, 1990.

Shirley Chisholm

Anderson, Delores Joan. "Black Women and Politics: Intersectionality of Race and Gender and the Transformative Production of Knowledge in Political Science." Ph.D. diss., The Union Institute, 2000.

Barnwell, Cherron Annette. "The Dialogics of Self in the Autobiographies of African-American Public Women: Ida B. Wells, Shirley Chisholm, Angela Davis and Anita Hill." Ph.D. diss., Howard University, 2002.

Brownmiller, Susan. *Shirley Chisholm*. Garden City, N.Y.: Doubleday, 1970.

Canas, Kathryn Anne. "Barbara Jordan, Shirley Chisholm, and Lani Guinier: Crafting Identification through the Rhetorical Interbraiding of Value." Ph.D. diss., University of Utah, 2002.

Chisholm, Shirley. *The Good Fight*. New York: Harper & Row, 1973.

———. *Unbought and Unbossed*. Boston: Houghton Mifflin, 1970.

Duffy, Susan, comp. *Shirley Chisholm: A Bibliography of Writings by and about Her*. Lanham, Md.: Rowman & Littlefield, 1988.

Haskins, James. *Fighting Shirley Chisholm*. Los Angeles: Dutton, 1975.

Hicks, Nancy. *The Honorable Shirley Chisholm: Congresswoman from Brooklyn*. New York: Lion, 1971.

Marshall-White, Eleanor. *Women: Catalysts for Change; Interpretive Biographies of Shirley St. Hill Chisholm, Sandra Day O'Connor, and Nancy Landon Kassebaum*. New York: Vantage, 1991.

Scheader, Catherine. *Shirley Chisholm: Teacher and Congresswoman.* Berkeley Heights, N.J.: Enslow, 1990.

"Shirley Anita Chisholm." In *Black Americans in Congress, 1870–1989.* Prepared under the direction of the Commission on the Bicentenary by the Office of the Historian, U.S. House of Representatives. Washington, D.C.: Government Printing Office, 1991.

"Shirley Anita Chisholm." In *Women in Congress, 1917–2006.* Prepared under the direction of the Committee on House Administration by the Office of History and Preservation, U.S. House of Representatives. Washington, D.C.: Government Printing Office, 2006.

I am sometimes left to dance across the cosmic floor alone, but I live by the faith that having the courage to take to the dance floor of life creates the possibility for a better world to spin into being.
    —Conrad Pegues

10. A Summing Up: Entanglements and Adjournments

Nothing is sadder than the waning dream of integration. This dream illuminated American life for the past several decades—. . .the progress in civil rights has not produced racial integration . . . Five decades after the Brown *ruling, black and white Americans do not live side by side, even when they share the same income levels. They do not go to the same schools. And when they do go to the same schools, they do not lead shared lives. . . . Maybe the health of a society is not measured by how integrated each institution within it is, but by how freely people can move between institutions. In a sick society, people are bound by one totalistic identity. In a healthy society, a person can live in a black neighborhood, send her kids to Catholic school, go to work in a lawyer's office, and meet every Wednesday with a feminist [womanist] book club. Multiply your homogenous communities and be fulfilled.*
    —David Brooks

Freedom takes on a different guise in each generation. Once upon a time, a crossing of the Canadian border was enough. Now, with cameras on city streets, the demise of habeas corpus, genocide in strife-ridden global communities, and the death-dealing "liberation" of oil-rich Middle Eastern countries, new definitions are needed. What does freedom mean if the planet dies? How shall we be free while others suffer?

We thought that we had it all figured out, but how could we have presumed that liberation would be easy? There is not much in the human realm that is! The rhetorical flourishes of activists and leaders who projected the ultimate outcomes of liberation theology included goals that seemed sappier than the ending of a 1940s musical. We envisioned a conclusion to strife and a finale where all God's children would dance into the sunset together. It just didn't happen, and perhaps it's for the best.

The first thing that happened to liberation is that the color line became a spectrum. Those caught in the snares of black/white discourse soon found themselves surrounded by "others" whose needs and dreams were as deeply rooted as their own. As the memories of plantations, whips, and chains faded, new cries for liberation were heard on the Mexican border, in Sudan, and in sweat shops in L.A. We thought that we were in control. We imagined an outcome that did not seem reasonable even to us, but once you imagine something it's hard to let it go.

We wanted a specified result that did not comport with the liberation that is seeded in the universe and in the human heart. We ignored a history replete with cautionary tales.

> For hundreds of thousands of years our ancestors lived in small bands. Surviving meant being able to distinguish between us—the people who will protect you—and them—the people who will kill you. Even today, people have a powerful drive to distinguish between us and them . . . no matter how arbitrary the basis of distinction—they will quickly begin discriminating against others they deem unlike themselves. People say they want to live in diverse integrated communities, but what they really want to do is live in homogenous ones, filled with people like themselves.[1]

So the first task in the quest for liberation is to tell the truth about who we are and what we really want and then add to this volatile mix a reality check about the universe(s) that we inhabit. No matter what social scientists say, we are connected to our choices and to one another because the world is not "out there"; it's here and there and everywhere.

Writer Sarah, could we wrap this up?

Sarah Nothing really ends, but if you need a stopping point, this one is as good as any.

Writer The meeting was amazing. At first, I thought that there were several meetings, but it was really one in several phases. I was always being surprised, especially when it came to the science aspect. When Einstein showed up in the early chapter, I thought that I would be swamped with technical language. Instead, it turns out that the story of the cosmos is one of emergence, creativity, and humility.

Sarah That's right. Despite the complexity, when we look up we engage the mystery of life and its origins in ways that should contribute to our humility and gratitude as a species. Only narrow scientific margins and vast amounts of grace allow our existence. When we view the heavens, the mysteries of dark matter, dark energy, and quantum entanglement, it should remind us that everything that matters is offered as gift and grace.

Writer We have not purchased the night stars; we can't put our air supply or sunshine on a credit card. The story of the cosmos has a moral: in the presence of such wonder, be humble, and be gentle with one another.

Sarah And hear the wisdom of the ages in the stories borne in your bodies and in the narratives bequeathed by forebears.

Writer I feel so honored to have heard from the elders.

Sarah Although you had a unique experience of presence, these stories are available to anyone. So much is revealed in the stories of their lives.

Writer Each ancestor or elder brought different areas of interest to my attention, which created a narrative tapestry. We began with story lines of the personal and communal journey. Each person reminded us that the universe also has a story that is grounded in connection. Our desires for a lasting liberation require that we reignite an interest in the common good. "All over the world, from every imaginable background and system of belief, people report that the trance of separation is being broken."[2]

What is certain is that liberation for all is different from liberation for one. The fragments of our striving are part of a wholeness that does not allow us to disappear into its comprehensiveness; rather, it is the totality of oneness that requires that we be grounded and embodied. Singer, scholar, and activist Bernice Johnson Reagon said that Negro spirituals recognized this wholeness but required the individual commitment of the word "I." One can shimmy out from under the responsibility of "we," but "I'm gonna let *my* little light shine" is a completely different thing. It is personal commitment publicly stated.

Liberation from a cosmological view must be grounded in the body in very specific ways because "the body is our earth; it is our most intimate connection with the planet. It has a gift to offer. It is alive and has an aspect to its nature that is always in the present."[3] It is the disassociation from body/earth that leads to an inability to hear one another. Our health and our well-being require that we embody our deepest and most heartfelt commitments.

Sarah I admit that I was expecting different stories from the elders, but there seemed to be a common theme. Did you notice it?

Writer Certainly. Their personal agendas did not predominate. The message is clear: we are integral parts of a wholeness that is not always apparent. If we are to fulfill the dreams of liberation that are articulated in each generation, we will have to shift from competition to cooperation. We must nurture a completely different mind-set that represents a shift in priorities. This is not just a strategy to accomplish goals for one marginalized community; it is a commitment to the liberation of all.

Sarah Now you're getting it. To fit into the story of an interconnected universe, the human community must relinquish the deadly myth of radical individualism. I call it deadly because it shrouds the real identity of humankind in a swirl of egocentric options that are not rooted in the story of the universe.

> Before the Big Bang our universe was compressed. Since then it has been expanding, but is has remained *entangled*. The findings of modern physics and the principles of perennial wisdom traditions are converging around this notion that everything is interconnected. And the spiritual values associated with those wisdom traditions are being validated by science as beneficial to individuals and society.[4]

Writer Liberation is the desire of all human beings. This common yearning is also what sustains the spark of hope that is passed down through the generations. My great grandmother didn't see freedom ring, but it didn't stop her from making crab cakes and sharing what she had. My father heard the dream articulated by Martin Luther King Jr. and walked in Selma toward the realization of the beloved community but did not arrive. Now it is my turn. I'm not certain that I know much more than my family members did about building a peaceful, interconnected society, but I do know that it will not come about because disparate racial/ethnic communities hoard their breakthroughs and throw rocks at those who are still banging on the door of inclusion.

Somehow God's diverse human community must share in the efforts to liberate one another. When we are all free, then perhaps the "beloved community" will come into sight.

Sarah "In order to move forward, we must reconnect with a wider and deeper wisdom, for it is clear that no single culture or perspective holds all the answers."[5]

Writer Just imagine the transformative power that would be unleashed if we connected the wisdom of the elders from all indigenous realms and their progeny, from Celtic to Sundancers, from Navajo Grandmothers to Irish storytellers, from Korean ritual dancers and hip-hop mime artists to African griots. I don't know about you, but I'm ready for the community-called-beloved to come into view. I am ready to work toward that end, beyond my own needs and in tandem with service to communities willing to journey toward reunion.

Sarah Amen. Meeting adjourned.

Notes

Preface
Epigraphs: Derrick Bell, *We Are Not Saved: The Elusive Quest for Racial Justice* (New York: Basic, 1987), xi. Judy Cannato, *Radical Amazement: Contemplative Lessons from Black Holes, Supernovas, and Other Wonders of the Universe* (Notre Dame, Ind.: Sorin, 2006), 26.

Chapter 1: Introduction
Epigraphs: Amy Shuman, *Other People's Stories: Entitlement Claims and the Critique of Empathy* (Champaign: University of Illinois Press, 2005). Fred Alan Wolf, *Mind into Matter: A New Alchemy of Science and Spirit* (Needham, Ma.: Moment Point, 2000).

1. Brief descriptions of the lives and legacies of the ancestors can be found at the beginning of each chapter.

2. The following quotation gives a brief synopsis of the "talented tenth." "The Negro race, like all races, is going to be saved by its exceptional men [*sic*]. The problem of education, then, among Negroes must first of all deal with the Talented Tenth; it is the problem of developing the Best of this race that they may guide the mass away from the contamination and death of the Worst, in their own and other races." W. E. B. DuBois, "The Talented Tenth," in *The Negro Problem: A Series of Articles by Representative Negroes of To-day*, ed. Booker T. Washington (New York: J. Pott, 1903).

3. Regarding the phrase "North American Indian," people native to the Americas have their own evolving naming process that is similar to the process African Diasporan people continue to experience. Some prefer to be called Native Americans; others, Indians or North American Indians. I use the phrase that I have been assured is respectful of their continuing process, with full awareness that self-naming takes a while in a postsubjugation reality.

4. Comedian and educator Bill Cosby has used the phrase "change the object of their desires" in recent years to describe one solution to the violence of young people today.

5. Will Coleman, "'Amen' and 'Ashe': African American Protestant Worship and Its West African Ancestor," *Cross Currents: Liturgy and Its Discontents* 52, no. 2 (2002): 158–64, quote at 159–60.

6. Ibid.

7. In the American judicial system, interlocutory rulings are temporary and intended to engage critical issues that affect an ongoing lawsuit.

8. Barbara A. Holmes, *Race and the Cosmos: An Invitation to View the World Differently* (Harrisburg, Pa: Trinity Press International, 2002), xvi.

9. Ibid., xvii.

10. I am grateful for the work of Hugh Lacey. His article entitled "Liberation Philosophy and Politics," *Cross Currents* 35, no. 2/3 (1985): 219–41, helped to shape some of the issues for consideration.

Chapter 2: Liberating the Idea of Freedom

Epigraphs: Traditional Negro spiritual written during the Civil War by emancipated black soldiers in the Union Army. Albert Einstein, *New York Post*, November 28, 1972.

1. Bishop Bettye Alston, "Ain't Ya Free Yet, Why Ain'y Ya Gone Free?" sermon given at New Day International Church, Memphis, Tenn., Spring 1999.

2. Albert Raboteau, "Afterword," in *An Unbroken Circle: Linking Ancient African Christianity to the African-American Experience,* ed. Paisuius Altschul (St Louis, Mo.: Brotherhood of St. Moses the Black, 1997), 162.

3. David Lantigua, "Freedom and the Dialogical Self," http://aporia.byu.edu/pdfs/lantigua-freedom_and_the_dialogical_self.pdf, p. 2.

4. Lantigua, "Freedom and the Dialogical Self."

5. Hugh Lacy, "Liberation: Philosophy and Politics," *Cross Currents* 35 (Summer/Fall 1985): 220, 235.

6. Breyten Breytenbach, "Imagine Africa," *Harpers Magazine*, May/June 2007, 15–18, quote at 16.

7. Dr. Billy Taylor is the Robert L. Jones Distinguished Professor of Music at East Carolina University in Greenville, North Carolina. The song "I Wish I Knew How It Would Feel to Be Free" was composed in 1954 and was sung by Nina Simone on the album *Silk and Soul.*

8. The origins of the song "We Shall Overcome" indicate a variety of sources. "[The] Lyrics [are] derived from Charles Tindley's gospel song 'I'll Overcome Some Day' (1900), and [the] opening and closing melody from the nineteenth-century spiritual 'No More Auction Block for Me' (a song that dates to before the Civil War). According to Professor Donnell King of Pellissippi State Technical Community College (in Knoxville, Tenn.), 'We Shall Overcome' was adapted from these gospel songs by Guy Carawan, Candy Carawan, and a couple of other people associated with the Highlander Research and Education Center, currently located near Knoxville, Tennessee. [Donnell King says] I have in my possession copies of the lyrics that include a brief history of the song, and a notation that royalties from the song go to support the Highlander Center." Eileen Southern, *The Music of Black Americans: A History*, 2nd ed. (New York: Norton, 1971), 546–47, 159–60.

9. Rufus Burrow, *James Cone and Black Liberation Theology* (Jefferson, N.C.: McFarland, 1994), 53.

10. Frank D. Macchia, "Jan Milič Lochman: A Tribute to My *Doktorvater*" *Pneuma: The Journal of the Society for Pentecostal Studies* 29, no. 1 (2007): 1–3, quote at 2.

11. Barbara Brown Taylor, *The Luminous Web: Essays on Science and Religion* (Cambridge, Mass.: Cowley, 2000), 90.

12. Brian Swimme, *The Universe Is a Green Dragon: A Cosmic Creation Story* (Rochester, Vt.: Bear, 2001), 28, 29.

13. John D. Norton, "Chasing a Beam of Light: Einstein's Most Famous Thought Experiment," http://www.pitt.edu/~jdnorton/goodies. Dr. Norton is a scholar in the Department of History and Philosophy of Science at the University of Pittsburgh.

14. Jennifer Derryberry, "The Beauty of Truth," *Science and Spirit: Inescapable Creativity* 13, no. 3 (May-June 2002): 54.

15. D. Bohm and B. Hiley, "On the Intuitive Understanding of Non-locality As Implied by Quantum Theory," *Foundations of Physics* 5 (1975): L 94.

Chapter 3: Law and Liberation

Epigraphs: Teilhard de Chardin, *Building the Earth*, trans. Noel Lindsey (Wilkes Barre, Pa.: Dimension, 1965), 33. Bishop Oscar Romero, *The Violence of Love*, tr. James R. Brockman (Maryknoll, N.Y.: Orbis, 2004).

1. Barbara Jordan, "The Universalization of the Philosophy or Ethic of Responsibility," speech given at University of Louisville Law School, Louisville, Ky., May 15, 1994.

2. Quoted in Carl T. Rowan, *Dream Makers, Dream Breakers; The Word of Justice Thurgood Marshall* (Boston: Little, Brown, 1993), 453–54.

3. William Shenston, *Essays on Men, Manners, and Things*, quoted in Brian Clegg, *The God Effect: Quantum Entanglement, Science's Strangest Phenomenon* (New York: St. Martin's, 2006), 1.

4. Barbara Jordan and Shelby Hearon, *Barbara Jordan: A Self-Portrait* (New York: Doubleday, 1979).

5. *Bush v. Gore* 121 S. Ct. at 530 (2004).

6. David A. Strauss, "*Bush v. Gore*: What Were They Thinking?" in *The Vote: Bush, Gore and the Supreme Court*, ed. Cass R. Sunstein and Richard A. Epstein (Chicago: University of Chicago Press, 2001).

7. Rafael Chodos, "Law as Dance, Theater, or Music: Legal Procedure and Ritual," *Cross Currents* 52, no. 2. (Summer 2002): 212–33, quote at 233.

8. Nikita Kruschev, speech in Yugoslavia, August 24, 1963.

9. Sarah Baxter, "Squalor of the Vet's Hospital Shocks U.S.," *Times Online*, March 4, 2007.

10. Copyright © Molly Ivins, "Habeas Corpus R.I.P. (1215–2006)," Truthdig.com, September 27, 2006.

11. Ivins, "Habeas Corpus R.I.P."

12. Kathleen Duffy, S.S.J., "The Texture of the Evolutionary Cosmos: Matter and Spirit in Teilhard de Chardin," in *Teilhard in the Twenty-First Century: The*

Emerging Spirit of Earth, ed. Arthur Fabel and Donald St. John (Maryknoll, N.Y.: Orbis, 2003), 143.

13. Pierre Teilhard de Chardin, *The Heart of Matter* (New York: Harcourt Brace, 1978), 16.

14. Constance Baker Motley, *Equal Justice under the Law: An Autobiography by Constance Baker Motley* (New York: Farrar, Strauss and Giroux, 1998), 245, 246.

15. I am indebted to the scholarship of Marcia Y. Riggs on issues of stereotype and myth.

Chapter 4: A Liberated and Luminous Darkness

Epigraphs: Vincent Harding, *There Is a River: The Black Struggle for Freedom in America* (New York: Harcourt Brace, 1993). Matthew Fox, *Creativity: Where the Divine and Human Meet* (New York, Putnam, 2002). Joanna Macy, quoted in *Creativity: Where the Divine and Human Meet* (New York, Putnam, 2002).

1. Rosa Parks and Gregory J. Reed, *Quiet Strength: The Faith, the Hope, and the Heart of a Woman Who Changed a Nation* (Grand Rapids, Mich.: Zondervan, 1994), 32, 88.

2. Parks and Reed, *Quiet Strength*, 91–93.

3. Barbara Ann Holmes, *Joy Unspeakable: Contemplative Practices of the Black Church* (Minneapolis: Fortress Press, 2004), 163.

4. Howard Thurman, *The Luminous Darkness: A Personal Interpretation of the Anatomy of Segregation and the Ground of Hope* (New York: Harper & Row, 1965), vii, viii.

Chapter 5: A Revolutionary Liberation

Epigraphs: Angela Davis, *African American Philosophers,* ed. George Yancy (New York: Routledge, 1998). James Cone, *For My People: Black Theology and the Black Church* (Maryknoll, N.Y.: Orbis, 1984).

1. Tubman's birth date is approximate because slaveholders did not always keep accurate records.

2. Sarah H. Bradford, *Harriet Tubman, The Moses of Her People* (Gloucester, Mass.: P. Smith, 1981); The American Experience Series, AE 7, 1st ed. (published under the title *Scenes in the Life of Harriet Tubman* by Sarah H. Bradford [Auburn, N.Y.: W. J. Moses, printer, 1869]).

3. Harriet was given a basket nickname, "Araminta," at birth. Dr Turner discusses the use of basket names in Gullah culture, but the use of African nicknames or names that reflect the circumstances of birth was common in slave communities. Turner says: "'The first scholar to make a serious study of the Gullah language was the late Dr. Lorenzo Turner, who published his findings in 1949. As a Black American, Dr. Turner was able to win the confidence of the Gullah people, and he revealed many aspects of their language that were previously unknown. Dr. Turner found that Gullah men and women all have African nicknames or 'basket names' in addition to their English names for official use; and he showed that the Gullah language, like other Atlantic Creoles, contains a substantial minority of vocabulary words borrowed directly from African substrate languages." Joseph A. Opala, *The Gullah: Rice, Slavery and the Sierra Leone–American Connection* (Freetown, Sierra Leone: USIS, 1987) (PM7875G80631987).

4. Malcolm X, *Chicago Defender*, November 28, 1962.
5. George Breitman, ed., *Malcolm X Speaks* (New York: Grove, 1965), 68–69.
6. Breitman, *Malcolm X Speaks*, 133.
7. Breitman, *Malcolm X Speaks*, 134.
8. Frederick Douglass, "The Significance of Emancipation in the West Indies," speech given in Canandaigua, New York, August 3, 1857; collected in pamphlet by author in *The Frederick Douglass Papers*, Series One: Speeches, Debates, and Interviews, vol. 3: 1833–63, ed. John Blassingame (1857; reprint, New Haven, Conn.: Yale University Press, 1985), 204. See also http://www.buildingequality.us/Quotes/Frederick_Douglass.htm. An earlier version of this quote is found in a letter to another abolitionist dated 1849. See Kim Bobo, Jackie Kendall, and Steve Max, eds., *Organizing for Social Change: A Manual for Activity in the 1990s* (Washington, D.C.: Seven Locks, 1991).
9. Quoted in Peniel E. Joseph, *Waiting 'Til the Midnight Hour: A Narrative History of Black Power in America* (New York: Henry Holt, 2006), 131.
10. Stokely Carmichael, "Berkeley Speech," in *Contemporary American Voices: Significant Speeches in American History, 1945–Present*, ed. James R. Andrews and David Zarefsky (New York: Longman, 1992), 100–107.
11. James Cone, "Martin, Malcolm, and Black Theology," in *The Quest for Liberation and Reconciliation: Essays in Honor of J. Deotis Roberts*, ed. Michael Battle (Louisville, Ky.: Westminster John Knox, 2005), 59.
12. Victor Anderson, *Beyond Ontological Blackness: An Essay on African American Religious and Cultural Criticism* (New York: Continuum, 1995), 117.
13. Michael Talbot, *The Holographic Universe* (New York: Harper Collins, 1991), 1.
14. Barbara A. Holmes, *Joy Unspeakable: Contemplative Practices of the Black Church* (Minneapolis: Fortress Press, 2004), 167.

Chapter 6: Liberated Bodies, Liberated Lives

Epigraphs: Arthur Ashe, *Days of Grace* (New York: Ballantine, 1994). Dorothee Soelle, *The Silent Cry: Mysticism and Resistance* (Minneapolis: Fortress Press, 2001), 112.
1. Audre Lorde, *Sister Outsider: Essays and Speeches by Audre Lorde* (Freedom, Calif.: Crossing, 1984, 1996), 142.
2. Fannie Lou Hamer, referenced in *This Little Light of Mine: The Life of Fannie Lou Hamer* by Key Mills (Lexington, Ky.: University of Kentucky Press, 2007).
3. "George Washington Carver Quotes" About.com: African American History, http://afroamhistory.about.com/od/georgewcarver/a/gwcarverquotes.htm.
4. Sandra Crouse Quinn and Stephan B. Thomas, "The National Negro Health Week, 1915-1951: A Descriptive Account," *Minority Health Today* 2(3): 44-49.
5. John Blake, "Thousands of Blacks March in Protest of Gay Marriage," *Atlanta Journal Constitution*, December 11, 2004.
6. Jasmyne Cannick, "Gays Lose Advocate with Death of Mrs. King," *New America Media*, February 3, 2006.
7. Peter Singer, *Animal Liberation*, rev. ed. (New York: Ecco, 2002).
8. Judy Ann Bigby, "The Challenge of Eliminating Disparities in Health," *Journal of General Internal Medicine* 17, no. 6 (2002): 489–90.

9. Cynthia Winton-Henry with Phil Porter, *What the Body Wants* (Kelowna, British Columbia: Northstone, 2004), 155.

10. Stephen Leahy, "Monsanto 'Seed Police' Scrutinize Farmers" (Inter Press Service), January 13, 2004.

11. Peter J. Gomes, "Black Christians and Homosexuality: The Pathology of a Permitted Prejudice," *African American Pulpit* 10, no. 3 (Summer 2007): 14–17, quote at 16.

12. Harriet A. Washington, *Medical Apartheid: The Dark History of Medical Experimentation on Black Americans from Colonial Times to the Present* (New York: Doubleday, 2006), 189–90.

13. Barbara Harris, quote from the founder of Children Requiring A Caring Kommunity, cited in Washington, *Medical Apartheid*, 189–90.

14. Robert Pollack, "DNA and Neshamah: Locating the Soul in the Age of Molecular Medicine," *Cross Currents* 53, no. 2 (Summer 2003): 231-47.

15. This is true whether one supports posthumanism or, like bioconservatives Francis Fukuyama or Leon Kass, one warns of its dangers. In fact, according to Jeffrey Mestoy, "the term posthuman is so much a part of the lexicon of biotechnology and artificial intelligence research" that neither Fukuyama, author of *Our Posthuman Future: Consequences of the Biotechnology Revolution*, nor Kass, author of *Life, Liberty and the Defense of Dignity: The Challenge for Bioethics*, found it necessary to define the phrase.

16. N. Katherine Hayles, *How We Became Posthuman: Virtual Bodies in Cybernetics, Literature and Informatics* (Chicago: University of Chicago Press, 1999), 3.

17. Nick Bostrum, "In Defense of Posthuman Dignity," *Bioethics* 19, no. 3 (2005): 202–14.

18. Ibid.

19. Troy Duster, "The Hidden Eugenic Potential of Germ-line Interventions," in *Designing Our Descendants: The Promises and Perils of Genetic Modifications*, ed. Audrey R. Chapman and Mark S. Frankel (Baltimore: John Hopkins University Press, 2003), 158, 159.

20. Barry Mehler suggests that the ideology and practices of eugenics remain with us in ways that should give us pause. Just a few years ago, a violence initiative coordinated by the National Institute of Mental Health considered procedures to "screen out and treat preventively violence prone individuals." See Barry Mehler, "In Genes We Trust," *Reform Judaism* (Winter 1994): 10–14, 77–79. Another violence initiative involved the joint participation of the Centers for Disease Control, the U.S. Justice Department, and the National Science Foundation. That study "called for more attention to biological and genetic factors in violent crime" (Mehler, "In Genes We Trust"). Even more troubling was the call for increased pharmaceutical research to identify groups with a higher propensity for violence. Meanwhile, eugenics research projects that reprised old but not forgotten ideas of inferiority continue to be well funded.

21. Marilyn E. Coors, *The Matrix: Charting an Ethic of Inheritable Genetic Modification* (New York: Rowman & Littlefield, 2002), 131, 132.

22. Coors, *The Matrix*. Although humans share most of their 32,000 genes, there are "500,000 gene components or single nucleotide polymorphisms (SNPs),

many of which are more common among people from one geographical region than from another." The medical community has more difficulty finding a transplant tissue match for African Americans because there are more antigen (protein) combinations on their cell surfaces than on the cells of white patients. Some of these antigens are very rare in the population at large. Some doctors give lower doses of narcotics to Asian patients because of the tendency toward an acute sensitivity to the effects of those drugs. Higher mortality from breast cancer among African American women—50 percent higher incidence before the age of thirty-five, as well as the highest incidence of premenopausal cancer. Zuni Native Americans have a unique mutation of cystic fibrosis. Genetic tests detect only the DF508 mutation prevalent in European descendants. The test detects over 90 percent of the mutations in whites but only 70 percent in blacks and less than 30 percent in Asians.

23. Consider the Canadian tuberculosis (TB) epidemic of the 1950s. All TB patients received the standard treatment: many months of triple-drug therapy, including a medication called Isoniazid. It turned out that a sizeable fraction of Canadian Innuits has a variant form of a liver enzyme that metabolized isoniazid so quickly that the drug was effectively used up before it could attack the tuberculosis bacteria. Many of the Innuits metabolized isoniazid much faster than the general population and thus fared poorly under what was an inadvertent two-drug regimen. Many succumbed to TB and the partly-treated, still-living tuberculosis bacteria themselves mutated into drug-resistant forms that went on to infect others in the general Canadian population. The significance of these results is that to ignore race under such circumstances is practically akin to withholding treatment. Sally Satel, "Medicine's Race Problem," *Policy Review* 110 (December 2001/January 2002): 49-59.

24. Ibid.

25. Quoted in George J. Annas, Lori B. Andrews, and Rosario M. Isasi, "Protecting the Endangered Human: Toward an International Treaty Prohibiting Cloning and Inheritable Alterations," *American Journal of Law and Medicine* 28, no. 2/3 (2002): 162.

26. Joseph Wood Krutch, "The Colloid and the Crystal," in *The Oxford Book of Essays*, ed. John Gross (Oxford: Oxford University Press, 1991), 447–48.

27. Leon Kass, *Life, Liberty, and the Defense of Dignity: The Challenge for Bioethics* (San Francisco: Encounter, 2002), 4.

28. Brian Swimme, *The Universe Is a Green Dragon: A Cosmic Creation Story* (Rochester, Vt.: Bea & Company, 2001), 115.

29. Swimme, *The Universe Is a Green Dragon*, 138, 139.

Chapter 7: Killing Me, Killing You, Killing Us!

Epigraphs: Angela Davis, "The Legacy of George Jackson," *Daily World*, August 25, 1971. James Baldwin, *The Price of the Ticket: Collected NonFiction, 1948-1985* (New York: St. Martin's, 1985), 686. Reneé Girard, *Things Hidden since the Foundation of the World* (London: Continuum International, 2003).

1. Ida B. Wells-Barnett, "This Awful Slaughter," in *Great Speeches by African Americans*, ed. James Daley (Mineola, N.Y.: Dover, 2006).

2. Interview with Amy Goodman, "A Conversation with Death Row Prisoner Stanley Tookie Williams from his San Quentin Cell," *Democracy Now: The War and Peace Report*, November 30, 2005.

3. Huey P. Newton, "Prison, Where Is Thy Victory?: January 3, 1970," in *The Huey P. Newton Reader*, ed. David Hilliard and Donald Weise (New York: Seven Stories, 2002).

4. William Stringfellow, *The Politics of Spirituality* (Philadelphia: Westminster, 1984), 70.

5. Vincent P. Franklin, *Living Our Stories, Telling Our Truths: Autobiography and the Making of African American Intellectual Tradition* (New York: Scribner, 1995).

6. Editorial written by Ida B. Wells-Barnett under the pen name Iola after the lynchings of three black grocers in Memphis. The article appeared in the *Free Speech and Headlight*, 1892.

7. Foreign Prisoner's Support System, December 13, 2005, 3:38 A.M. ET. http://www.phaseloop.com/foreignprisoners/case-stanley-william.html

8. Amy Goodman, "A Conversation with Death Row Prisoner Stanley Tookie Williams from his San Quentin Cell," *Democracy Now!*, November 30, 2005, http://www.democracynow.org/article.pl?sid=05/11/30/153247 (accessed June 8, 2008).

9. Wells-Barnett, "This Awful Slaughter," 100.

10. Troy Duster, "Pattern, Purpose and Race in the Drug War: The Crisis of Credibility in Criminal Justice," in *Crack in America: Demon Drugs and Social Justice*, ed. Craig Reinarman and Harry G. Levine (Berkeley: University of California Press, 1997), 260–87.

11. Documentation of counterintelligence came in a report issued in 1976 entitled *The FBI's Covert Program to Destroy the Black Panther Party*. The investigation was authorized by the U.S. Senate Select Committee to Study Government Operations.

12. Available at http://www.blackpanther.org/TenPoint.htm (accessed June 8, 2008).

13. Joanna Russ, "Power and Helplessness in the Women's Movement," in *Magic Mommas, Trembling Sisters, Puritans and Perverts: Feminist Essays* (Trumansburg, N.Y.: Crossing, 1985); also cited by Janice McLane, "The Boundaries of Victim Life," in *Interrogating Ethics: Embodying the Good in Merleau-Ponty*, ed. James Hatley, Janice McLane, and Christian Diehm (Pittsburgh: Duquesne University Press, 2006), 135.

14. Michael Talbot, "The Universe as a Hologram," http://twm.co.nz/hologram.html (accessed June 8, 2008). See also Talbot's discussion of this experiment in *Mysticism and the New Science* (New York: Penguin Putnam, 1981, 1993), 141–43.

15. Poetic excerpts appear in Roger Guenveur Smith, "The Journey: From Stage to Film," a summary of the making of the film *A Huey P. Newton Story*, Luna Ray Films, 2002.

16. Smith, "The Journey: From Stage to Film."

17. Gary Gunderson and Larry Pray, *Leading Causes of Life* (Memphis, Tenn.: Cause Leader, 2006).

Chapter 8: Liberation and the Art of Creative Imagination

Epigraphs: Gordon Parks, interview with Phil Ponce "Half Past Autumn," published as a NewsHour transcript, Online NewsHour, January 6, 1998. http://www.pbs.org/newshour/bb/entertainment/jan-june98/gordon_1-6.html. Tupac Shakur, referenced in ThugLifeArmy: A Total News Source for Hip-Hop Culture. http://www.thuglifearmy.com/news/?id=4389.

1. Tupac Shakur, unattributed quotation referenced in *Tupac Shakur Legacy* by Jamal Joseph (New York: Atria, 2006).
2. Phillis Wheatley in a letter written to Reverend Samson Occum on February 11, 1774 (Connecticut Gazette, March 11, 1774).
3. Langston Hughes, "The Negro Artist and the Racial Mountain," *Nation* (1926).
4. Gwendolyn Brooks, "The Chicago Picasso," *In the Mecca: Poems by Gwendolyn Brooks* (New York, Harper & Row, 1968).
5. Phillis Wheatley, "On Being Brought from Africa" *Poems on Various Subjects Religious and Moral* by Phillis Wheatley, Negro Servant to Mr. John Wheatley of Boston, New England, London, printed for A. BELL bookseller, Algate and sold by Messrs. COX and BERRY, King Street, Boston, June 12, 1773. MDCCLXXIII.
6. Henry Louis Gates, "Mister Jefferson and the Trials of Phillis Wheatley" (lecture delivered on March 22, 2002, at the Ronald Reagan International Trade Center, Washington, D.C.), 16, 17.
7. Gates, "Mister Jefferson," n. 21, 22.
8. Henry Miller, "Art or Propaganda?: Major Twentieth-Century African American Approaches to Dramatic Theory" (unpublished paper, the Graduate School and University Center of the City University of New York, Spring 1996), 5.
9. Miller, "Art or Propaganda?," 6.
10. Miller, "Art or Propaganda?," 6.
11. Alain Locke, "Negro Art," *The Crisis* 21 (June 1921): 55–56.
12. Peter Paris, *The Spirituality of African Peoples: The Search for a Common Moral Discourse* (Minneapolis: Fortress Press, 1995). 147-48.
13. Matthew Fox, *Creativity: Where the Divine and Human Meet* (New York, Putnam, 2002), 35.

Chapter 9: Beyond Mountaintops

Epigraphs: Octavia Butler, "On Racism" for U.N. World Conference on Racism, National Public Radio, August 20, 2001. http://www.npr.org/templates/story/story.php?storyId=5245686. Harry Belafonte, interview with Sarah Rugh van Gelder, "Freedom Sings" *Yes! Magazine* (Spring 2002). http://www.yesmagazine.org/article.asp?ID=487.

1. Martin Luther King Jr., presidential address, Southern Christian Leadership Conference, Atlanta, August 16, 1967.
2. Martin Luther King Jr., "Loving Your Enemies" sermon delivered at Dexter Avenue Baptist Church, Montgomery, Alabama, November 17, 1957.
3. This excerpt is from the Last Will and Testament of Mary McLeod Bethune to Bethune Cookman College. The testament is inscribed on a sculpture erected in Lincoln Park, Washington, D.C., in 1974 to honor her. Originally published in *Ebony*, August 1955.

4. Shirley Chisholm, "The Black Woman in Contemporary America," speech given at a conference on black women in America at the University of Missouri, Kansas City, June 17, 1974; also cited in *Great Speeches by African Americans*, ed. James Daley (Mineola, N.Y.: Dover, 2006), 134.

5. Rosemary Freeney Harding and Rachel Elizabeth Harding, "Hospitality, Haints, and Healing: A Southern African American Meaning of Religions," in *Deeper Shades of Purple: Womanism in Religion and Society*, ed. Stacey M. Floyd-Thomas (New York: New York University Press, 2006), 100.

6. A version of this conversation was included in published lectures offered during the Princeton Youth Lectures, Seattle, Washington, 2006.

7. Sally A. Brown, "When Lament Shapes the Sermon," in *Lament: Reclaiming Practices in Pulpit, Pew, and Public Square*, ed. Sally A. Brown and Patrick D. Miller (Louisville, Ky.: Westminster John Knox, 2005), 35.

8. Dorothee Soelle, *The Silent Cry: Mysticism and Resistance* (Minneapolis: Fortress Press, 2001), 77–93.

9. Kevin Sharpe, *Sleuthing the Divine: The Nexus of Science and Spirit* (Minneapolis: Fortress Press, 2000), 76.

10. bell hooks, *Teaching Community: A Pedagogy of Hope* (New York: Routledge, 2003), xiv.

11. Barbara Ann Holmes, *Race and the Cosmos: An Invitation to View the World Differently* (Harrisburg, Pa.: Trinity Press International, 2002).

12. John Polkinghorne, *Science and Theology: An Introduction* (Minneapolis: Fortress Press, 1998), 72-73.

Chapter 10: A Summing Up

Epigraphs: Conrad Pegues, *Spirited: Affirming the Soul and Black Gay/Lesbian Identity*. Eds. G. Winston James and Lisa C. Moore (Washington D.C.: RedBone, 2006), 278. David Brooks, "Maybe Integration Shouldn't Define Us," *Memphis Commercial Appeal*, July 7, 2007, A11.

1. David Brooks, "Maybe Integration Shouldn't Define Us," *Memphis Commercial Appeal*, July 7, 2007, A11.

2. Arjuna Ardagh, "Cultivating Translucence," *Shift: The Transformative Power of Learning* 8 (September-November 2005): 28–30, quote at 30.

3. Carrie Gray, "Before Knowing Being," *Shift: The Transformative Power of Learning* 8 (September-November 2005): 34–37, quote at 34.

4. *The 2007 Shift Report: Evidence of a World Transforming* (Petaluma, Calif.: Institute of Noetic Sciences, 2007), 73.

5. *The 2007 Shift Report*, 66.

Index

African American theology, 92–93
African Diaspora
   and ancestors, 8–9
   and communal goals, 65–66
   cosmology, 9
   and economic stability, 106
   on freedom/liberation, 19, 177
   and grief, 178
   and justice, 135
   and liberation leaders, 4
agency and liberation, 21, 101
AIDS. *See* HIV/AIDS
ancestors, defined, 8–9
Anderson, Victor, 19, 92–93
Aristotle, 19
arts and music, 145–61
   and community, 156
   and cosmology, 157–60
   and prophesy, 14
Ashe, Arthur, 99
Aspect, Alain, 138–39

Baldwin, James, 125
Belafonte, Harry, 167
Bereano, Philip, 118
Bethune, Mary McLeod
   background, 168–69
   imagined dialogue with King, DuBois, Chisholm, 168–69
beyondness. *See* other side
bioconservativism, 116
biogenetics, 102, 114–15, 116–17, 120

bisexuals. *See* GLBT community
Black Panther Party, 83–84, 127–28, 135–37
   ten points, 136
black power, 89
Bohm, David, 95
bondage, 26, 152
Bretenbach, Breyten, 22
Brooks, David, 187
Brooks, Gwendolyn
   background, 147–48
   imagined dialogue with Shakur, Langston, Wheatley, 149–61
Brown, James, 149
Brown, John, 70, 82
Bukovsky, Vladimir, 48
Burrow, Rufus, 26
Bush, Barbara, 159
Bush, George W.
   on illegal imprisonment, 47–48
   on Thomas appointment, 53
   on torture, 48
Bush v. Gore, 40–41
Butler, Octavia, 167

Cannon, Katie G., 19
Carmichael, Stokely, 89, 90–91
Carver, George Washington, 2
   background, 101
   imaginary dialogue with Lorde and Hamer, 102–21
Center for Genetics and Society, 115

203

Chisholm, Shirley
  background, 170
  imagined dialogue with King, DuBois, Bethune, 170
Civil Rights movement
  and black power, 89
  and dream language, 21–22
  and law, 13
  leaders, 4, 168
  music in, 22
  and racism, 24
  and timelessness, 72
  women in, 174–75
COINTELPRO. *See* Counterintelligence Program
community
  dissolution of, 156
  focus on good in, 71
  health of, 109–10, 114
  immersion in, 27, 190
  and pursuit of freedom, 30, 106
  solidarity and liberation, 23–34
Cone, James H., 19, 23, 74, 92
conformity, 72
consciousness
  black, 92
  and freedom/liberation, 27, 95, 138
  of nation, 29
Constitution, U.S., 39, 42–43
consumerism, 6, 65, 70
Cosby, Bill, 180
cosmology
  of African Diaspora, 9
  and art, 157–60
  and connection, 3, 113, 139
  and freedom, 13, 83, 179–80
  and healing, 178
  of liberation, 18, 19–23, 24–25, 28–29, 30, 93–94, 106, 121, 172, 180
Counterintelligence Program (COINTELPRO), 135
Crips, 127, 132

Davis, Angela, 19, 81, 125
death penalty, 44
difference, 51, 93, 119
discrimination and solidarity, 24
diversity, 87, 106
domination systems, 19, 21, 23, 28–29

Douglass, Frederick, 89–90, 155
dream language, 21–22
DuBois, W. E. B., 1, 2, 126, 155
  background, 169–70
  imagined dialogue with King, Hamer, Bethune, Chisholm, 169–70
Dussel, Enrique, 19

economic security, 106, 107
Einstein, Albert, 17
  imagined dialogue with Sarah/Writer, 31–34
elders
  on moral imagination, 24
  questions for, 11–15
  wisdom of, 10–11, 108, 191
  See also specific individuals
El-Shabazz, El-Hajj Malik. *See* Malcolm X
equality, 38
eugenics, 198n18
Eurocentrism, 63
Evers, Medgar, 129

FBI. *See* Federal Bureau of Investigation (FBI)
fear
  of difference, 51
  overcoming, 86
  and terror, 5, 51, 131
Federal Bureau of Investigation (FBI), 135
Fox, Matthew, 158
freedom
  and agency, 21, 101
  and arts, 149
  as choice, 25
  and community, 30, 106
  in cosmological context, 13, 83, 179–80
  defined, 18, 20–21
  embodiment of, 14
  and health, 108
  and justice, 29
  and law, 13
  legal codification, 38
  meaning of, 12
  spiritual characteristics of, 13
  work of, 4
  See also liberation

Fukuyama, Francis, 116

gangs, 132
Garvey, Marcus, 1, 2
Gates, Henry Louis, 151–52
gay-lesbian-bisexual-transgender community. See GLBT community
gender
  bias, 104
  politics, 104, 170
genetics, 114–15, 118–19
germline modification, 114–15, 118–20
Girard, Reneé, 125
GLBT community, 23–24, 107–8, 111–12
globalization, 91
God
  depiction of, 74–75
  and humankind, 95
  liberating, 21
  liberation in, 71, 148
  modeling freedom, 25
Gomes, Peter J., 111
Gonzáles, Justo, 19
guerilla tactics, 85
Gullah people, 3, 196n3
Gutiérrez, Gustavo, 19

habeas corpus, 47
Hamer, Fannie Lou, 1
  background, 100–101
  imaginary dialogue with King and DuBois, 170–73
  imaginary dialogue with Lorde and Carver, 109–21
Hampton, Fred, 140
Harding, Vincent, 59
Harlem Renaissance, 154
Hayles, N. Katherine, 114–15
health, 105, 108–10, 114, 116, 135, 187. *See also* medical ethics
HIV/AIDS, 107, 179
Hoover, J. Edgar, 135
Houston, Charles Hamilton, 37
Huggins, Johnny, 140
Hughes, Langston (James), 2, 148
  background, 147
  imagined dialogue with Shakur, Wheatley, Brooks, 149–61

identity
  formation, 20, 62
  sexual, 111–12
incarceration rates, 134
inferiority, 19, 91, 110, 117–19, 138, 156, 198 n.18
integration, 44, 63, 81, 187
Isasi-Díaz, Ada María, 19
Ivins, Molly, 47–48

Johnson, James Weldon, 148
Jones, William R., 19
Jordan, Barbara, 2
  background, 35–36
  on constitutional codification, 39
  imaged dialogue with Marshall, 40–54
  as legal expert, 35
justice
  and cosmos/community, 27
  and freedom, 29
  and power, 92
  racial, 108, 135
  work of, 4

Kass, Leon, 116, 121
King, Bernice, 107
King, Coretta Scott, 2, 107–8, 168
King, Martin Luther, Jr., 1, 2, 89, 190
  background, 167–68
  as civil rights spokesperson, 66, 168
  death, 129, 168
  imagined dialogue with Hamer, DuBois, Bethune, Chisholm, 84
  on injustice, 108
  on nature of revolutions, 84
  Thurman as advisor to, 62
Kirk-Duggan, Cheryl, 23
Kwok Pui-Lan, 19

lament, 175–78
land and liberation, 88, 110
law
  defined, 45
  inequity in application, 43
  and liberation/freedom, 13, 35–58
leadership, 15, 73, 170, 174
lesbians. *See* GLBT community
liberation
  and agency, 21, 101
  and arts, 145–61

and communal solidarity, 23–34
and connection, 30
and consciousness, 95, 138
cosmologies of, 18, 19–23, 24–25, 28–29, 30, 93–94, 106, 172, 180
defined, 20–21
as egalitarian, 4
and faith traditions, 13
in God, 71
and law, 13, 35–58
and leadership, 15
and love, 53–54
of movement, 93
philosophies of, 12, 19–23
processive view, 26–27, 29
roots of, 18
and survival, 6
and violence, 14, 125–42
and wisdom, 6–7
work of, 4
See also freedom
liberation leaders, 4. *See also* specific individuals
liberation projects
in African American context, 4
status, 2
and values, 14
liberation theology
black, emergence of, 22–23, 26
on empowerment, 25
Liouzo, Viola, 129
Little, Malcolm. *See* Malcolm X
Locke, Alain, 155
Long, Charles W., 19
Lorde, Audre
background, 99–100
imaginary dialogue with Hamer and Carver, 102–21
poetry of, 102
love and liberation, 53–54
luminous darkness, 75–76
lynching, 126, 130–31, 133–34

Machado, Daisy, 19
Macy, Joanna, 59
Magna Carta, 47, 48
Malcolm X. 1, 2
background, 82–83
death, 129
imaginary dialogue with Tubman, 83–95

market economy
and oppressed people, 5
Marshall, Thurgood
background, 37–38
on death penalty, 38
imagined dialogue with Jordan, 40–54
as legal expert, 35
medical ethics, 102–3, 104, 116–17, 118
Mehler, Barry, 198n18
Meredith, James, 89
Mini, Vuyisile, 160
misogyny, 152
moral imagination, 24, 31
moral responsibility, 86, 87
Motley, Constance Baker, 53
music
in Civil Rights movement, 22
rap, 152–53, 156, 159
on women, 152–53
See also arts and music

National Association for the Advancement of Colored People (NAACP), 37, 60, 126, 169
National Council of Negro Women, 169
nationalism, 81
National Minority Health Foundation, 105
National Negro Health Week, 105
Nelson, Roger, 31–32
Newton, Huey P.
background, 127–28
imagined dialogue with Wells-Barnett, 133–41
Nightingale, R., 131
9/11 attacks. *See* September 11, 2001 terrorist attacks
Nixon, Richard M., 36
nonlocality concept, 33
nonviolent resistance. *See* resistance
Norvell, Aubrey James, 89

oppression
and control, 137
and economics, 5, 70
hold of, 22
and inclusion, 6, 29
inner/outer, 21, 137–38

and law, 13
and liberation, 18, 19, 74
overcoming, 62, 64, 131
and revolution, 86
types, 173
other side
   accessibility of, 8
   boundaries, 171
   defined, 2
   and time-space, 4

Paris, Peter J., 157
Parks, Gordon, 145
Parks, Rosa, 1
   background, 60
   imagined dialogue with Thurman, 62–76
Pegues, Conrad, 187
Pinn, Anthony B., 19
Plato, 19
Poe, Edgar Allan, 59, 115
Polkinghorne, John, 32
Pollack, Robert, 114
posthumanism, 114–21
poverty and law, 13
power, 67–68
process theology, 26–27
prophesy and art, 14
protests, 29
Public Health Service, U. S., 105
Raboteau, Albert, 19

race
   burden of, 99
   relations in twenty-first century, 5–6
   and violence, 129
racism
   after Civil Rights movement, 24
   created by people, 99
   and medical ethics, 102–3, 104, 116–17
   reverse, 91
Reagan, Ronald, 159
Reagon, Bernice Johnson, 189
reconciliation, 139
religion
   and domination systems, 5
   support of liberation in, 13–14
reproductive rights, 113, 116
resistance, 38, 73–74, 84, 89–90, 131

reverse racism, 91
revolution, 86–96
   in black consciousness, 92
   nature of, 84, 88, 90
   and oppression, 86
   scientific, 95
   of values, 88, 96
rhetoric vs. prophecy, 5
Riggs, Marcia Y., 19, 23
rights
   individual, 29
   reproductive, 113
rituals, 178–179
Romero, Oscar, 35, 129
Ross, Araminta. *See* Tubman, Harriet

Samuel, 7
Sarah (martyr), 7
   imagined dialogues with Writer, 8–10, 24–34, 54–55, 76–77, 96, 121–22, 141–42, 161, 181, 188–91
Saul, King, 7
Seale, Bobby, 127, 135
September 11, 2001 terrorist attacks, 49–50, 176
sexual politics, 106, 108
Shakur, Tupac
   background, 145–46
   imagined dialogue with Langston, Wheatley, Brooks, 149–61
Shurman, Amy, 1
Simone, Nina, 22
Singer, Peter, 109
slavery
   and art, 154
   revolts, 70
   in twenty-first century, 5, 6
   voluntary, 66
SNCC. *See* Student Nonviolent Coordinating Committee (SNCC)
Soelle, Dorothee, 99
solidarity and discrimination, 24. *See also* community
spirit realm. *See* other side
spirituality, African, 29
stem cell research, 118
storytelling, 1, 22
Strauss, David A., 40
Stringfellow, William, 128
Student Nonviolent Coordinating

Index   207

Committee (SNCC), 101
survival and liberation, 6

Talbot, Michael, 94
Taylor, Barbara Brown, 27
Taylor, Billy, 22
Teilhard de Chardin, Pierre, 35, 52, 53–54
terror/terrorism, 5, 49, 51, 112–13, 131. *See also* September 11, 2001 terrorist attacks
theology. *See* African American theology; liberation theology; process theology
Thomas, Clarence, 53
Thurman, Howard
background, 60
imagined dialogue with Parks, 62–76
Till, Emmett, 72, 129
time-space dimensions
boundaries, 171–72
and other side, 4, 10
and universe, 17
torture, 47–48
Townes, Emilie, 19
transgender/transsexuals. *See* GLBT community
transhumanism, 115–16
truth, 19, 29, 44–45, 94–95, 153, 188
Tubman, Harriet
background, 82
imaginary dialogue with Malcolm X, 83–95
Turner, Nat, 70

underground railroad, 85, 91
unity, 173
universe
story of, 30–31
and wholeness, 33–34, 190

values in liberation projects, 14
violence
conditions for, 132
of cross, 137
and liberation, 14, 125–42
and nonlocality, 138, 140
ordinariness of, 129
and race, 129
and reason, 134

and resistance, 84, 89
voter registration, 100–101

Washington, Booker T., 105, 126, 169
Wells-Barnett, Ida B., 1, 70
background, 126
imagined dialogue with Newton, 133–41
imagined dialogue with Williams, 130–33
West, Cornel, 19
Wheatley, Phillis
background, 146–47
imagined dialogue with Shakur, Langston, Brooks, 149–61
poetry, 147
white supremacy, 89–90
wholeness, 28, 33–34, 121, 173, 189–90
Williams, Stanley Tookie
background, 126–27
imagined dialogue with Wells-Barnett, 130–33
Willliams, Raymond, 127
wisdom
of ancestors, 9–10
and liberation, 6–7
Wolf, Fred Alan, 1
women
in Civil Rights movement, 174–75
in music, 152–53
rights of, 169
Writer
described, 2
imaged dialogues with Sarah, 8–10, 24–34, 54–55, 76–77, 96, 121–22, 141–42, 161, 181, 188–91

x-ray dosages, 116–17

208   Liberation and the Cosmos